Dugard

Applied
Fluid
Mechanics

Applied
Fluid
Mechanics

Robert L. Mott
University of Dayton

CHARLES E. MERRILL PUBLISHING COMPANY
A Bell & Howell Company
Columbus, Ohio

International Standard Book Number: 0-675-09144-6

Library of Congress Catalog Card Number: 73-183356

7 8 — 78

Printed in the United States of America

Preface

The objective of this book is to present the basic principles of fluid mechanics and the application of these principles to practical, applied problems. Primary emphasis is on the topics of fluid statics, flow of fluids in pipes, open channel flow, flow measurement, and forces developed by fluids in motion.

Applications are shown in the fields of Mechanical Engineering Technology including fluid power, Chemical Engineering Technology including flow in processing systems, and Civil Engineering Technology as applied to fluid storage and distribution systems and open channel flow. The book is directed to anyone in a technical field where the ability to apply the principles of fluid mechanics is desirable.

Those using the book are expected to have an understanding of algebra, trigonometry, and physics mechanics. After completing the book, the student should have a sufficiently firm foundation in the field to continue learning if he desires. However, it is not specifically intended to prepare a student for more advanced courses in fluid mechanics. Other applied courses, such as fluid power, could be taught following this course, or this book could be used to teach selected fluid mechanics topics within such a course.

The approach used in the book encourages the student to become involved in the learning of the principles of fluid mechanics at three levels:
1. The understanding of concepts,
2. The recognition of the logical approach to problem solutions,
3. The ability to perform the details required in the solutions.

This multilevel approach should build the student's confidence in his ability to solve problems.

Concepts are presented in clear language and illustrated by reference to physical systems with which the reader should be familiar. An intuitive justification as well as a mathematical basis is given for each concept.

The methods of solution to many types of complex problems are presented in step-by-step procedures. The importance of recognizing the relationships among what is known, what is to be found, and the choice of a solution procedure is emphasized.

Many practical problems in fluid mechanics require relatively long solution procedures. It has been my experience that students often have difficulty in accomplishing the details of the solution. For this reason, each example problem its worked in complete detail including the manipulation of units in equations. In the more complex examples, a programmed instruction format is used in which the student is asked to perform a small segment of the solution before being shown the correct result. The programs are of the linear type in which one panel presents a concept and then either poses a question or asks that a certain operation be performed. The following panel gives the correct result and the details of how it was obtained. The program then continues.

Dimensional units in the English gravitational system are consistent throughout the book. Graphs and tables of data are presented in the units in which problems are to be solved, thus streamlining the required computations. The importance of consistent units in equations is strongly emphasized. Conversions are presented between the English units and the International System of Units (SI), the metric system, and other convenient units.

I would like to thank Professor L. Duke Golden for encouraging me to undertake the writing of the book; and Professor Jesse H. Wilder, my department chairman, for his cooperation and advice during the preparation of the manuscript. To all others who helped in any way, I express my appreciation.

Robert L. Mott
March, 1972

Contents

1

Introduction, Units, and Calculations

Why Study Fluid Mechanics?

When you turn on a faucet, water is delivered to you by a distribution system composed of pumps, valves, and pipes. The source of the water may be a storage tank, reservoir, river, lake, or well. The flow of water from its source to your faucet is controlled by the principles of fluid mechanics. An understanding of these principles is necessary in order to properly select the size and type of pumps and pipes, to design the storage tanks, to select flow control valves, and to monitor the performance of the system.

The performance of an automated manufacturing machine, which is controlled by fluid power systems, is dependent on the flow of the hydraulic oil and the pressure at the actuators. A typical system is shown in Fig. 1-1. The greater the pressure of the oil in a cylinder, the

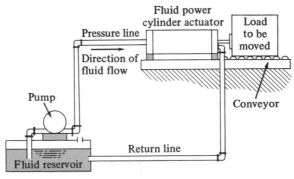

FIGURE 1-1

1

greater the force it can exert. The greater the flow rate of oil into the cylinder, the faster it will move. You will learn how to analyze such a system using the material in this book.

A buoy marking a boating channel seems like a very simple device, and it is. However, the material from which the buoy is made and its geometry must be specified according to the laws of buoyancy and stability of floating bodies which you will learn in Chapter 3 of this book.

In your automobile, fuel is delivered to the carburetor from the tank by a fuel pump. How much power is being supplied by the engine to drive the pump? The material in Chapter 5 will help in calculating this.

A highway sign giving directions to motorists may have to withstand high winds. In order to determine the forces exerted on the sign due to the winds, the principle of impulse-momentum as presented in Chapter 13 must be understood.

These are just a few of the many practical problems likely to be encountered which require an understanding of the principles of fluid mechanics for their solution. The objective of this book is to help you solve these kinds of problems. Included in each chapter are problems representing situations from many fields of technology. Your ability to solve these problems will be a measure of how well the objective of this book has been accomplished.

1–2   What Is Fluid Mechanics?

Fluid mechanics is the study of the behavior of fluids either at rest or in motion. Fluids can be either gases or liquids. You are probably familiar with the physical difference between gases and liquids, as exhibited by air and water, but for the study of fluid mechanics it is convenient to classify fluids according to their compressibility.

    ✳ Gases are very readily compressible.
    Liquids are only slightly compressible.

For many practical applications these definitions are sufficient. The method of solution of a problem often depends on whether the fluid is compressible or not.

In this book it will be assumed that liquids are totally incompressible unless otherwise noted. This means that the density of liquids will be considered constant no matter how much pressure is applied. Unless the change in pressure in a particular situation is several thousand pounds per square inch, this assumption will not cause a significant error in calculations. Most of the material in this book will deal with liquids.

Since gases are very readily compressible we should take the change in density with pressure into account. This greatly increases the complexity of the analysis of a problem since the principles of thermodynamics must be used to produce accurate results when even moderate changes in pressure occur. In some sections of this book the compressibility of all fluids will be carefully accounted for. In other sections, however, gases will be considered incompressible in order to obtain a quick rough estimate of certain situations. Although this is not strictly accurate, there are times when an estimate is sufficient.

1-3   **The English Gravitational Unit System**

In any technical work the units in which physical properties are measured must be stated. A system of units specifies the units of the basic quantities of length, time, force, and mass. The units of other terms are then derived from these.

The English gravitational unit system will be used in this book. The units for the basic quantities are:

$$\begin{cases} \text{length} = \text{foot (ft)} \\ \text{time} = \text{second (sec)} \\ \text{force} = \text{pound (lb)} \\ \text{mass} = \text{slug (lb-sec}^2\text{/ft)} \end{cases}$$

Probably the most difficult of these units to understand is the slug since we are familiar with measuring in terms of pounds, seconds, and feet. It may help to note the relationship between force and mass from physics.

$$F = ma$$

where a is acceleration having the units of ft/sec^2. Therefore, the derived units for mass are

$$m = \frac{F}{a} = \frac{\text{lb}}{\text{ft/sec}^2} = \frac{\text{lb-sec}^2}{\text{ft}} = \text{slug}$$

This means that we may use either slugs or lb-sec^2/ft for the unit of mass. In fact, some calculations in this book require that you be able to use both or to convert from one to the other.

1-4   **Consistent Units in an Equation**

The analyses required in fluid mechanics involve the algebraic manipulation of several terms. The equations are often complex and it is extremely important that the results be dimensionally correct. That is, they must have the proper units. Indeed, answers will have the wrong numerical value if the units in the equation are not consistent.

A simple straightforward procedure called unit cancellation will insure proper units in any kind of calculation, not only in fluid mechanics but also in virtually all your technical work. The six steps of the procedure are listed below.

unit analysis

Unit Cancellation Procedure

1. Solve the equation algebraically for the desired term.
2. Decide on the proper units for the result.
3. Substitute known values including units.
4. Cancel units which appear in both the numerator and denominator of any term.
5. Use conversion factors to eliminate unwanted units and obtain the proper units as decided in Step 2.
6. Perform the calculation.

This procedure, properly executed, will work for any equation. It is really very simple but some practice may be required to use it. We are going to borrow some material from elementary physics with which you should be familiar to illustrate the method. However, the best way to learn how to do something is to do it. The following example problems are presented in a form called programmed instruction. You will be guided through the problems in a step-by-step fashion with your participation required at each step.

To proceed with the program, you should cover all material below the heading "Programmed Example Problems" using a heavy piece of paper. You should have a blank piece of paper handy on which to perform the requested operations. Then successively uncover one panel at a time down to the heavy line which runs across the page. The first panel presents a problem and asks you to perform some operation or to answer a question. After doing what is asked, uncover the next panel which will contain information that you can use to check your result. Then continue with the next panel and so on through the program.

Remember, the purpose of this is to help you learn how to get correct answers using the unit cancellation method. You may want to refer to the table of conversion factors in the appendix.

Programmed Example Problems

Example Problem 1–1.   Imagine you are traveling in a car at a constant speed of 70 miles per hour. How many minutes would it take to travel 5 miles?

For the solution, use the equation

$$s = vt$$

where s is the distance traveled, v is the speed, and t is the time. Using the unit cancellation procedure outlined above, what is the first thing to do?

The first step is to solve for the desired term. Since you were asked to find time you should have written

$$t = \frac{s}{v}$$

Now perform Step 2 of the procedure.

Step 2 is to decide on the proper units for the result, in this case time. From the problem statement the proper unit is minutes. If no specification had been given for units you could choose any acceptable time unit such as seconds or hours.

Proceed to Step 3.

The result should look something like this.

$$t = \frac{s}{v} = \frac{5 \text{ mi}}{70 \text{ mi/hr}}$$

Notice that the units for v are written as a fraction, mi/hr, and not as mph as is often seen. For the purpose of cancellation it is not convenient to have the units in the form of a compound fraction as we have above. To clear this to a simple fraction, write it in the form

$$t = \frac{\dfrac{5 \text{ mi}}{1}}{\dfrac{70 \text{ mi}}{\text{hr}}}$$

This can be reduced to

$$t = \frac{5 \text{ mi-hr}}{70 \text{ mi}}$$

After some practice, equations may be written in this form directly.

Now perform Step 4 of the procedure.

The result should now look like this.

$$t = \frac{5 \text{ mi-hr}}{70 \text{ mi}}$$

Here is an illustration that units can be cancelled just as numbers can if they appear in both the numerator and denominator of a term in an equation.

Now do Step 5.

The answer looks like this.

$$t = \frac{5 \text{ mi-hr}}{70 \text{ mi}} \cdot \frac{60 \text{ min}}{1 \text{ hr}}$$

The equation in the preceding panel showed the result for time in hours after the mile units were cancelled. Although hours is an acceptable time unit our desired unit is minutes as determined in Step 2. Thus the conversion factor 60 min/1 hr is required.

How did we know to multiply by 60 instead of dividing?

The units determine this. Our objective in using the conversion factor was to eliminate the hour unit and obtain the minute unit. Since the unwanted hour unit was in the numerator of the original equation the hour unit in the conversion factor must be in the denominator in order to cancel.

Now that we have the time unit of minutes we can proceed with Step 6.

The correct answer is $t = 4.29$ min. A slide rule should be used for every calculation no matter how simple it appears to be. As explained in Section 1–5 many problems in this book involve a great number of terms and to perform every calculation by hand would take too much time. By practicing on the simpler problems you will gain confidence in the slide rule results.

Example Problem 1–2. Figure 1–2 shows a fluid power actuator pushing a 100-pound box on a roller conveyor. Assuming there is no friction, calculate the force which must be exerted by the actuator to give the box a forward acceleration of 50 in./sec^2.

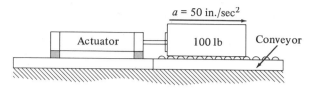

FIGURE 1–2

From physics you learned that force is proportional to acceleration. That is, $F = ma$. Also, mass and weight are related by $m = w/g$, where m is mass, w is weight, and g is the acceleration due to gravity. (Use $g = 32.2 \text{ ft/sec}^2$.)

Using the unit cancellation procedure to solve the problem, do the first step.

The result is

$$F = ma = \frac{wa}{g}$$

Proceed to Step 2.

No units are specified for the desired result so we are free to choose any force unit. Assuming you use pounds, proceed to Step 3.

The answer looks like this:

$$F = \frac{wa}{g} = \frac{(100 \text{ lb})(50 \text{ in./sec}^2)}{32.2 \text{ ft/sec}^2}$$

Remember to put in the units. This can be rewritten to clear the compound fractions of units as follows:

$$F = \frac{wa}{g} = 100 \text{ lb} \cdot \frac{50 \text{ in.}}{\text{sec}^2} \cdot \frac{\text{sec}^2}{32.2 \text{ ft}}$$

Proceed to Step 4.

You should now have

$$F = \frac{wa}{g} = 100 \text{ lb} \cdot \frac{50 \text{ in.}}{\cancel{\text{sec}^2}} \cdot \frac{\cancel{\text{sec}^2}}{32.2 \text{ ft}}$$

We are not finished yet since we have the inches and feet units remaining in addition to the desired unit of pounds. Therefore, do Step 5 now.

The result should be this:

$$F = \frac{wa}{g} = 100 \text{ lb} \cdot \frac{50 \cancel{\text{in.}}}{\cancel{\text{sec}^2}} \cdot \frac{\cancel{\text{sec}^2}}{32.2 \cancel{\text{ft}}} \cdot \frac{\cancel{\text{ft}}}{12 \cancel{\text{in.}}}$$

Since only the pound unit is left we can do Step 6.

The correct answer is $F = 12.9$ lb.

In performing the calculations from the previous panel the equation may appear to be too long and drawn out. If so, it may be helpful to rewrite it as follows keeping only the lb unit with the numbers.

$$F = \frac{(100)(50)}{(32.2)(12)} \text{ lb} = 12.9 \text{ lb}$$

Example Problem 1–3. Calculate the time in seconds required for a rock to fall 30.0 feet into a well if it starts from rest.

The equation governing this situation is

$$s = \tfrac{1}{2} g t^2$$

where s is the distance, t is the time, and g is the acceleration due to gravity (32.2 ft/sec²).

Perform the first four steps of the unit cancellation procedure before looking at the next panel.

So far you should have:
Step 1.

$$s = \tfrac{1}{2} g t^2$$

$$t = \sqrt{\frac{2s}{g}}$$

Step 2.

The desired units are seconds.

Step 3.

$$t = \sqrt{\frac{(2)(30 \text{ ft})}{32.2 \text{ ft/sec}^2}} = \sqrt{\frac{(2)(30 \text{ ft})}{1} \cdot \frac{\text{sec}^2}{32.2 \text{ ft}}}$$

Step 4.

$$t = \sqrt{\frac{(2)(30) \text{ ft-sec}^2}{32.2 \text{ ft}}}$$

Notice that the units are under the square root sign. Now proceed to Step 5.

There are no conversion factors required. Remembering that

$$\sqrt{\text{sec}^2} = \text{sec}$$

the result of Step 5 should be

$$t = \sqrt{\frac{(2)(30)}{32.2}} \text{ sec}$$

Using a slide rule for Step 6 we would get

$$t = 1.37 \text{ sec}$$

This concludes the programmed instruction.

===

You should now have an understanding of how to solve problems using unit cancellation. The problems after the next section can be used for additional practice.

1–5   **Calculations**

Solving problems in fluid mechanics requires many calculations. Listed below are seven typical calculations which are parts of problems found in later chapters of this book.

$$p = \frac{(28.5)(14.7)}{29.92} \tag{1–1}$$

$$N = \frac{(11.2)(0.275)(1.63)}{0.00032} \tag{1–2}$$

$$H = \frac{(0.0185)(2250)(17.2)^2}{(0.63)(2)(32.2)} \tag{1–3}$$

$$p = \frac{56.1}{144}\left[\frac{(8.2)^2 - (23.9)^2}{(2)(32.2)}\right] \tag{1–4}$$

$$H = \frac{(32)(4.9 \times 10^{-3})(189)(2.2)}{(62.4)(1.26)^2} \tag{1–5}$$

$$D = \frac{(0.021)(1280)(4.3)^2}{(\pi)^2(32.2)(14.3)} \tag{1–6}$$

$$D = (30.3)^{0.2} \tag{1–7}$$

These are shown here to emphasize the importance of using a slide rule for calculations. To perform the above calculations by long multiplication and division would take an exorbitant amount of time.

A common reason for not using a slide rule is lack of confidence in the results. Practice is the best solution for this problem. By using the slide rule for all calculations, even simple ones, your confidence in the results will grow. Then when more complex calculations are to be done the advantage of speed can be gained by using a slide rule.

The total procedure for performing calculations such as those in Eqs. (1–1) to (1–7) actually can be separated into two parts; finding the numbers and locating the position of the decimal point. The slide rule gives the numbers quickly and with sufficient accuracy for most technical calculations. The decimal point location can be found quickly by estimation.

The typical slide rule calculations required are:

Multiplication and division
Squaring and square root
Raising a number to an odd power [Eq. (1–7)]

Since the arrangement of scales varies on slide rules of different manufacturers, it is recommended that you refer to the manual for your own rule for instructions on the above operations. A simple procedure for estimating the decimal point location is described below.

Decimal Point Location by Estimation.   Equation (1–1) is considered first.

$$p = \frac{(28.5)(14.7)}{29.92}$$

The slide rule gives the sequence of numbers for the result as 140. The remaining problem is to decide if the final answer is 0.140, 1.40, 14.0, 140, 1400 or something else involving the numbers 140. The estimation procedure can be set up in a step-by-step manner as shown below.

Decimal Point Estimation Procedure

1. Round off numbers to easily manipulated values.
2. Express very large or very small numbers in terms of powers of 10.
3. Cancel powers of 10 if possible.
4. Cancel rounded-off numbers if possible.
5. Mentally (or on scratch paper) estimate the magnitude of the result.

Use the programmed instruction routine to illustrate this procedure. Cover all the material below the double line and then proceed one panel at a time.

Programmed Example Problems

Example Problem 1–4.   We are working with the problem

$$p = \frac{(28.5)(14.7)}{29.92}$$

for which the slide rule has been used to give the sequence of numbers in the answer as 140. Perform Step 1 of the decimal point estimation procedure before looking at the next panel.

The result should look like this.

$$p \cong \frac{(30)(15)}{30}$$

The symbol, $\cong$, means "is approximately equal to," since the value of p is only being estimated.

You can be very coarse in rounding at this point because the magnitude of the result needs to be estimated only within a factor of 10 in order to place the decimal point.

In this simple case you can neglect Steps 2 and 3 of the procedure and go on to Step 4.

Now you should have

$$p \cong \frac{(\cancel{30})(15)}{(\cancel{30})}$$

This means that the result is about 15 and therefore, using the numbers from the slide rule, the final answer is $p = 14.0$.

Now read the next panel for a more difficult problem.

Example Problem 1–5.   For Eq. (1–2),

$$N = \frac{(11.2)(0.275)(1.63)}{0.00032}$$

the slide rule gives the sequence of numbers in the answer as 157. Where should the decimal point be placed?

Perform Step 1 of the procedure.

This is one possible answer:

$$N \cong \frac{(10)(0.3)(2)}{0.0003}$$

This is much simpler than the original problem. Now do Step 2.

Writing the numbers in terms of powers of 10 gives

$$N \cong \frac{(10)(3 \times 10^{-1})(2)}{3 \times 10^{-4}}$$

Go on to Step 3.

The problem is now simplified to

$$N \cong \frac{(\cancel{10})(3 \times \cancel{10^{-1}})(2)}{3 \times 10^{-4}} \cong \frac{(3)(2)}{3 \times 10^{-4}}$$

Now do Steps 4 and 5.

Finally we have

$$N \cong \frac{(\cancel{3})(2)}{\cancel{3} \times 10^{-4}} \cong 2 \times 10^4$$

as an estimate of the magnitude of N. The final answer is

$$N = 1.57 \times 10^4 = 15,700$$

Example Problem 1–6. Equation (1–3) is

$$H = \frac{(0.0185)(2250)(17.2)^2}{(0.63)(2)(32.2)}$$

Notice that one of the numbers is to be squared. This result can be estimated too. First, however, begin with Step 1 of the procedure.

Here is one possible form of the rounded-off equation.

$$H \cong \frac{(0.02)(2000)(20)^2}{(1)(2)(30)}$$

Now do Step 2.

The result is

$$H \cong \frac{(2 \times 10^{-2})(2 \times 10^3)(2 \times 10)^2}{(2)(3 \times 10)}$$

Now take care of that number to be squared.

Since $(2 \times 10)^2 = 4 \times 10^2$ we now have

$$H \cong \frac{(2 \times 10^{-2})(2 \times 10^3)(4 \times 10^2)}{(2)(3 \times 10)}$$

Go on to Step 3.

We can eliminate many powers of 10 here.

$$H \cong \frac{(2 \times \cancel{10^{-2}})(2 \times \cancel{10^3})(4 \times 10^2)}{(2)(3 \times \cancel{10})}$$

$$H \cong \frac{(2)(2)(4)(10^2)}{(2)(3)}$$

Steps 4 and 5 should now be easy.

Performing the cancellation gives

$$H \cong \frac{(\cancel{2})(2)(4)(10^2)}{(\cancel{2})(3)} \cong \frac{(8)(10^2)}{3}$$

That is about 3×10^2 or 300. Using a slide rule gives the actual sequence of numbers as 304. Therefore, $H = 304$.

The solutions for Eqs. (1–4), (1–5), and (1–6) are worked out completely on the next three panels. As a check of your ability to perform decimal point estimation and slide rule operation, work each problem yourself before looking at the solution.

Example Problem 1–7.   Solution for Eq. (1–4).

$$p = \frac{56.1}{144}\left[\frac{(8.2)^2 - (23.9)^2}{(2)(32.2)}\right]$$

$$p \cong \frac{50}{150}\left[\frac{(10)^2 - (20)^2}{(2)(30)}\right]$$

$$p \cong \frac{5 \times 10}{1.5 \times 10^2}\left[\frac{100 - 400}{(2)(3 \times 10)}\right]$$

$$p \cong \frac{5 \times \cancel{10}}{1.5 \times \cancel{10^2}}\left[\frac{-3 \times \cancel{10^2}}{(2)(3 \times \cancel{10})}\right]$$

$$p \cong \frac{(5)(-\cancel{3})}{(1.5)(2)(\cancel{3})} \cong \frac{-5}{3} \cong -2$$

The actual answer is $p = -3.05$.
   The solution for Eq. (1–5) follows in the next panel.

Example Problem 1–8.   Solution for Eq. (1–5).

$$H = \frac{(32)(4.9 \times 10^{-3})(189)(2.2)}{(62.4)(1.26)^2}$$

$$H \cong \frac{(30)(5 \times 10^{-3})(200)(2)}{(50)(1)^2}$$

$$H \cong \frac{(3 \times \cancel{10})(\cancel{5} \times 10^{-3})(2 \times 10^2)(2)}{(\cancel{5} \times \cancel{10})} \cong 12 \times 10^{-1} \cong 1.2$$

The actual answer is $H = 0.658$. Notice that this is closer to the estimated answer of 1.2 than either 0.0658 or 6.58. So even a coarse estimate is sufficient.
   The solution for Eq. (1–6) follows in the next panel.

Example Problem 1–9.   Solution for Eq. (1–6).

$$D = \frac{(0.021)(1280)(4.3)^2}{(\pi)^2(32.2)(14.3)}$$

$$D \cong \frac{(0.02)(1000)(5)^2}{(3)^2(30)(15)} \cong \frac{(0.02)(1000)(25)}{(9)(30)(15)}$$

$$D \cong \frac{(0.02)(1000)(30)}{(10)(30)(20)} \cong \frac{(\cancel{2} \times 10^{-2})(\cancel{10^3})(\cancel{3} \times 10)}{(\cancel{10})(\cancel{3} \times \cancel{10})(2 \times \cancel{10})}$$

$$D \cong 1 \times 10^{-1} \cong 0.1$$

The actual answer is $D = 0.109$.

This concludes the programmed instruction for this section. You should now be able to quickly estimate the magnitude of a calculation for locating the decimal point.

The problems below are for extra practice on consistent units, slide rule calculations, and estimation of the decimal point location in a calculation. Those on conversion factors, slide rule, and decimal point location are patterned after actual calculations you will find in later parts of this book.

PRACTICE PROBLEMS

Conversion factors

1-1.   Convert 136 inches to feet.

1 2.   Convert 592 square inches to square feet.

1-3.   Convert 874 cubic inches to cubic feet.

1-4.   Convert 4.2 square feet to square inches.

1-5.   Convert 25 gallons to cubic feet.

1-6.   Convert 3.25 cubic feet to cubic inches.

1-7.   An automobile is moving at 50 mph. Calculate its speed in ft/sec.

1-8.   Convert a length of 572 centimeters to feet.

1-9.   Convert a distance of 4.3 kilometers to feet.

1-10.   Convert a length of 12.3 centimeters to feet.

1-11.   Convert a length of 8.6 kilometers to feet.

1-12.   Convert a volume of 4760 cubic centimeters to cubic feet.

1-13.   Convert a volume of 7390 cubic centimeters to gallons.

1-14.   Convert a volume of 175 cubic centimeters to cubic feet.

1-15.   Convert 32 centimeters per second to ft/sec.

1-16.   Convert 300 centimeters per minute to ft/sec.

Consistent units in an equation.

A body moving with constant velocity obeys the relationship: $s = vt$; where s = distance, v = velocity, and t = time.

1–17.   A car travels 1000 feet in 14 seconds. Calculate its average speed in miles per hour.

1–18.   In an attempt at a land speed record a car travels a mile in 5.7 seconds. Calculate its average speed in miles per hour.

A body starting from rest with constant acceleration moves according to the relationship: $s = \frac{1}{2} at^2$; where s = distance, a = acceleration, and t = time.

1–19.   If a body moves 3.2 kilometers in 4.7 minutes while moving with constant acceleration, calculate the acceleration in ft/sec².

1–20.   An object is dropped from a height of 53 inches. Neglecting air resistance, how long would it take for the body to strike the ground? Use $a = g = 32.2$ ft/sec².

The formula for kinetic energy is: $KE = \frac{1}{2} mv^2$; where m is mass and v is velocity.

1–21.   Calculate the kinetic energy in ft-lb of a 1-slug mass if it has a velocity of 4 ft/sec.

1–22.   Calculate the kinetic energy in ft-lb of an 8000-pound truck moving 10 miles per hour.

1–23.   Calculate the kinetic energy in ft-lb of a 150-pound box moving on a conveyor at 20 ft/sec.

1–24.   Calculate the mass of a body in slugs if it has a kinetic energy of 15 ft-lb when moving at 2.2 ft/sec.

1–25.   Calculate the weight of a body in pounds if it has a kinetic energy of 38.6 ft-lb when moving at 19.5 miles per hour.

1–26.   Calculate the velocity in ft/sec of a 30-pound object if it has a kinetic energy of 10 ft-lb.

1–27.   Calculate the velocity in ft/sec of a 6-ounce body if it has a kinetic energy of 30 in.-oz.

One measure of a baseball pitcher's performance is his earned run average or ERA. It is the average number of earned runs he has allowed if all the innings he has pitched were converted to equivalent nine inning games. Therefore, the units for ERA are runs per game.

1–28.   If a certain pitcher has allowed 39 runs during 141 innings, calculate his ERA.

1-29.   A pitcher has an ERA of 3.12 runs/game and has pitched 150 innings. How many earned runs has he allowed?

1-30.   A pitcher has an ERA of 2.79 runs/game and has allowed 40 earned runs. How many innings has he pitched?

1-31.   A pitcher has allowed 49 earned runs during 123 innings. Calculate his ERA.

Slide rule and decimal point location. Calculate the following.

1-32. $\dfrac{(85)(49)}{33}$

1-33. $\dfrac{(22.6)(3.84)}{32.2}$

1-34. $\dfrac{(1848)(0.292)}{4.86}$

1-35. $\dfrac{(30.22)(14.7)}{29.92}$

1-36. $(46.5)(62.4)(0.83)$

1-37. $\dfrac{(19.7)(56.1)(0.56)}{0.92}$

1-38. $\dfrac{0.0418}{0.0005}$

1-39. $\dfrac{\pi(3.5)^2}{4}$

1-40. $\dfrac{(26.6)(\pi)(0.75)^2}{4}$

1-41. $\dfrac{\pi[(4)^2 - (3)^2]}{(4)(144)}$

1-42. $\dfrac{0.86\,(12.62 - 2.39)^2}{64.4}$

1-43. $\dfrac{(31.4)^2 - (9.26)^2}{(2)(32.2)}$

1-44. $\dfrac{(7.15)^2 - (15.4)^2}{(2)(32.2)}$

1-45. $\dfrac{(0.0165)(4050)(14.3)^2}{(0.625)(64.4)}$

1-46. $\dfrac{(8)(755)(0.62)^2}{(\pi)^2(32.2)(16.3)}$

1-47. $\left[1 - \dfrac{0.0491}{0.1965}\right]^2 \dfrac{(5.72)^2}{64.4}$

1-48. $\dfrac{(10.8)(0.167)(1.94)}{6.3 \times 10^{-3}}$

1-49. $\dfrac{(20.7)(0.333)(1.65)}{3.4 \times 10^{-4}}$

1-50. $\dfrac{(4)(0.56)(1.86)}{(\pi)(0.904)(6.8 \times 10^{-6})}$

1-51. $\dfrac{(32)(6.8 \times 10^{-5})(2849)(9.86)}{(56.2)(0.333)^2}$

1-52. $\sqrt{(64.4)(1.28)}$

1-53. $\sqrt{\dfrac{(2)(32.2)(9.3)}{12}}$

1-54. $\sqrt{\dfrac{(0.094)(32.2)(2)(18.6)}{62.4}}$

1-55. $\sqrt{\dfrac{(64.4)(9.8)(12.6)(16)}{(12)(15)}}$

1-56. $\sqrt{\dfrac{(16)(64.4)(150 - 93.4)(144)}{(15)(72.2)}}$

1-57. $(0.69)(0.785)\left[\dfrac{(64.4)(12 + 22.9)}{\left[1 - \dfrac{0.785}{3.14}\right]^2}\right]^{1/2}$

1-58. $\dfrac{14.8}{\cos 35°}\left[\dfrac{32.2}{(2)[14.8 \tan 35° - (-2)]}\right]^{1/2}$

1-59. $10.29 \sin 60°$ 1-60. $5.63 \sin 50°$

1-61. $2.39 \cos 30°$ 1-62. $15.98 \cos 40°$

1-63. $4.73 \tan 40°$ 1-64. $16.8 \tan 55°$

The following calculations are typical of those found in certain pipe line problems. Remember, in general terms, $(n)^{0.2} = (n)^{1/5}$ and if $(n)^{1/5} = b$, then $n = b^5$.

1-65. $(1.072)^{0.2}$ 1-66. $(1.296)^{0.2}$

1-67. $(5.74)^{0.2}$ 1-68. $(14.5)^{0.2}$

1-69. $(30.8)^{0.2}$ 1-70. $(850)^{0.2}$

1-71. $(0.0433)^{0.2}$ 1-72. $(0.123)^{0.2}$

1-73. $(0.493)^{0.2}$ 1-74. $(0.625)^{0.2}$

1-75. $(0.942)^{0.2}$

2

Basic Fluid Properties
and Glossary of Terms

2–1   **Introduction**

The behavior of a fluid either at rest or in motion is dependent upon and indicated by the properties of the fluid. This chapter presents the definitions, symbols, and units for the basic fluid properties of mass, weight, density, specific weight, and specific gravity. You will need to understand these properties before you can proceed to the material presented later in this book.

Also presented in this chapter is a glossary of terms which are developed and used later. They need not be studied now but are listed here for reference at any time so you will not have to search through the book for a definition.

Unfortunately, there is not universal agreement on the symbols used for fluid properties and other parameters encountered in the study of fluid mechanics. Those used in this book are widely used. However, because you are likely to see different symbols in other books and articles, it is important for you to understand the basic definitions.

In the following sections the symbol and units (in the English gravitational unit system) for each term are shown at the left margin.

2–2   **Basic Fluid Properties**

m
slug or
lb-sec^2/ft

Mass is the property of a body of fluid that is a measure of its inertia or resistance to a change in motion.

19

w
lb

Weight is the amount that a body weighs; that is, the force with which a body is attracted toward the earth by gravitation. It is related to mass m, and the acceleration due to gravity g, by the relation, $w = mg$. In this book we will use $g = 32.2$ ft/sec^2.

ρ (rho)
slugs/ft^3
or lb-sec^2/ft^4

Density is the amount of mass per unit volume of a substance. Therefore, $\boxed{\rho = m/V}$ where V is the volume of the substance having a mass m.

γ (gamma)
lb/ft^3

Specific weight is the amount of weight per unit volume of a substance. Therefore, $\boxed{\gamma = w/V}$ where V is the volume of the substance having a weight w.

sg
unitless

Specific gravity can be defined in terms of either density or specific weight as in (a) and (b) below.

(a) *Specific gravity* is the ratio of the density of a substance to the density of water at 4°C.
(b) *Specific gravity* is the ratio of the specific weight of a substance to the specific weight of water at 4°C.

These definitions for specific gravity can be shown mathematically as

$$\text{sg} = \frac{\gamma_s}{\gamma_w \ @ \ 4°C} = \frac{\rho_s}{\rho_w \ @ \ 4°C}$$

where the subscript s refers to the substance whose specific gravity is being determined and the subscript w refers to water. The properties of water at 4°C are constant, having the values shown below.

$$\gamma_w \ @ \ 4°C = 62.4 \ \text{lb/ft}^3$$
$$\rho_w \ @ \ 4°C = 1.94 \ \text{slugs/ft}^3$$

Therefore, the mathematical definition of specific gravity can be written

$$\boxed{\text{sg} = \frac{\gamma_s}{62.4 \ \text{lb/ft}^3} = \frac{\rho_s}{1.94 \ \text{slugs/ft}^3}}$$

This definition holds regardless of the temperature at which the specific gravity is being determined.

The properties of fluids vary with temperature. In general, the density (and therefore, the specific weight and specific gravity) decreases with increasing temperature. The properties of water are listed in the appendix at various temperatures. For other liquids the properties are shown at a few selected temperatures.

Relation Between Density and Specific Weight. Quite often the specific weight of a substance must be found when its density is known and vice versa. The conversion from one to the other can be made by the following equation:

$$\gamma = \rho g$$

where g is the acceleration due to gravity, 32.2 ft/sec². This equation can be justified by referring to the definitions of density and specific gravity and by using the equation relating mass to weight, $w = mg$.

The definition of specific weight is

$$\gamma = \frac{w}{V}$$

By multiplying both the numerator and the denominator of this equation by g we obtain

$$\gamma = \frac{wg}{Vg}$$

But $m = w/g$. Therefore,

$$\gamma = \frac{mg}{V}$$

Since $\rho = m/V$,

$$\gamma = \rho g$$

The following problems illustrate the definitions of the basic fluid properties presented above and the relationships between the various properties. The mathematical definitions shown above in boxes are used in the solution of the problems. It is important to insure that the units of the desired property are correct. These problems also illustrate the unit cancellation procedure discussed in Chapter 1. Try the problems before looking at the solution in a manner similar to the programmed instruction used in Chapter 1.

EXAMPLE PROBLEMS

Example Problem 2–1.   A pint of water weighs 1.04 pounds. Find its mass.

Solution.   Since $w = mg$, the mass is

$$m = \frac{w}{g} = \frac{1.04 \text{ lb}}{32.2 \text{ ft/sec}^2} = \frac{1.04 \text{ lb-sec}^2}{32.2 \text{ ft}}$$
$$m = 0.032 \text{ lb-sec}^2/\text{ft} = 0.032 \text{ slugs}$$

Remember that the units of slugs and lb-sec^2/ft are the same.

Example Problem 2–2.   One cubic foot of mercury has a mass of 26.3 slugs. Find its weight.

Solution.

$$w = mg = 26.3 \text{ slugs} \cdot 32.2 \text{ ft/sec}^2 = 849 \text{ slug-ft/sec}^2$$

This is correct but the units may seem confusing since weight is normally thought of as being in pounds. The units of mass may be rewritten as lb-sec^2/ft.

$$w = mg = 26.3 \frac{\text{lb-sec}^2}{\text{ft}} \cdot \frac{32.2 \text{ ft}}{\text{sec}^2} = 849 \text{ lb}$$

Example Problem 2–3.   A pint of water at 40°F weighs 1.04 lb. Find (a) its specific weight in lb/ft^3, (b) its density in slugs/ft^3, and (c) its specific gravity.

Solution.   (a) By definition, specific weight is

$$\gamma = \frac{w}{V} = \frac{1.04 \text{ lb}}{1 \text{ pint}} \cdot \frac{8 \text{ pts}}{\text{gal}} \cdot \frac{7.48 \text{ gal}}{\text{ft}^3} = 62.4 \text{ lb/ft}^3$$

(b) Density is found from either

$$\rho = m/V \text{ or } \rho = \gamma/g$$

Using $\rho = \gamma/g$,

$$\rho = \frac{62.4 \text{ lb}}{\text{ft}^3} \cdot \frac{\text{sec}^2}{32.2 \text{ ft}} = 1.94 \frac{\text{lb-sec}^2}{\text{ft}^4} = 1.94 \frac{\text{slugs}}{\text{ft}^3}$$

(c) Specific gravity

$$sg = \frac{\gamma_s}{62.4 \text{ lb/ft}^3} = \frac{62.4 \text{ lb/ft}^3}{62.4 \text{ lb/ft}^3} = 1.00 \text{ (unitless)}$$

Example Problem 2–4.   Oil for an industrial hydraulic system has a specific gravity of 0.90 when its temperature is 100°F. Calculate its specific weight and its density.

Solution.

$$(sg)_o = \frac{\gamma_o}{62.4 \text{ lb/ft}^3} = \frac{\rho_o}{1.94 \text{ slugs/ft}^3}$$

In this case $(sg)_o = 0.90$, and the subscript o refers to the properties of the oil. Then,

$$\gamma_o = (sg)_o(62.4 \text{ lb/ft}^3) = (0.90)(62.4 \text{ lb/ft}^3)$$
$$\gamma_o = 56.2 \text{ lb/ft}^3$$
$$\rho_o = (sg)_o(1.94 \text{ slugs/ft}^3) = (0.90)(1.94 \text{ slugs/ft}^3)$$
$$\rho_o = 1.75 \text{ slugs/ft}^3$$

PRACTICE PROBLEMS

Note: A table of conversion factors is located in the appendix.

2–1. Calculate the mass of a gallon of oil if it weighs 7.8 pounds.

2–2. Calculate the mass of a cubic foot of gasoline if it weighs 42.0 pounds.

2–3. Calculate the weight of a cubic foot of kerosene if it has a mass of 1.58 slugs.

2–4. Calculate the weight of a gallon of water if it has a mass of 0.258 slugs.

2–5. The specific gravity of benzene is 0.88. Calculate its specific weight and its density.

2–6. Air at 59°F and standard atmospheric pressure has a specific weight of 0.0765 lb/ft³. Calculate its density.

2–7. Carbon dioxide has a density of 0.00381 slugs/ft³ at 32°F. Calculate its specific weight.

2–8. A certain medium lubricating oil has a specific weight of 56.4 lb/ft³ at 40°F and 54.0 lb/ft³ at 120°F. Calculate its specific gravity at each temperature.

2–9. At 212°F mercury has a specific weight of 830 lb/ft³. What volume of the mercury would weigh 500 pounds?

2–10. One gallon of a certain fuel oil weighs 7.50 pounds. Calculate its specific weight, its density, and its specific gravity.

2–11. Glycerine has a specific gravity of 1.26. How much would 50 gallons of glycerine weigh?

2-12. The fuel tank of an automobile holds 25.0 gallons. If it is full of gasoline having a density of 1.32 slugs/ft³, calculate the weight of the gasoline.

2-13. The density of muriatic acid is 1.20 grams per cubic centimeter (gm/cm³). Calculate its density in slugs/ft³, its specific weight in lb/ft³, and its specific gravity. (Note that specific gravity and density in grams per cubic centimeter are numerically equal.)

2-14. Liquid ammonia has a specific gravity of 0.89. Calculate the volume in cubic centimeters which would weigh 5.0 pounds.

2-15. Vinegar has a density of 1.08 gm/cm³. Calculate its specific weight in lb/ft³.

2-16. Alcohol has a specific gravity of 0.79. Calculate its density both in slugs/ft³ and gm/cm³.

2-17. A cylindrical container is 6 inches in diameter and weighs 0.50 pounds when empty. When filled to a depth of 8.0 inches with a certain oil it weighs 7.95 pounds. Calculate the specific gravity of the oil.

2-18. A storage vessel for gasoline (sg = 0.68) is a vertical cylinder, 30 feet in diameter. If it is filled to a depth of 22 feet, calculate how many gallons are in the tank and how much the gasoline weighs.

2-19. How many gallons of mercury (sg = 13.60) would weigh the same as five gallons of castor oil which has a specific weight of 60.3 lb/ft³?

2-20. A rock has a specific gravity of 2.32 and a volume of 8.64 cubic inches. How much does it weigh?

2-3   Glossary of Terms

This section lists the symbols, units, and definitions of several terms which are developed throughout the book. They are arranged alphabetically according to the name of the term.

p_{abs} *Absolute pressure* is measured relative to a perfect
psia   vacuum.

                A *barometer* is a device for indicating the magnitude of atmospheric pressure.

Buoyancy is the tendency for a fluid to exert a supporting force on a body placed in the fluid.

F_b
lb

The *buoyant force* is the supporting force exerted by a fluid on a body placed in the fluid and is equal to the weight of the fluid displaced by the body.

cb

The *center of buoyancy* is the point through which the buoyant force acts. It is at the centroid of the displaced volume of fluid.

The *center of pressure* is that point on an area at which the entire force due to fluid pressure can be considered to act.

The *centroid of an area* is the point at which the area would be balanced if suspended from that point.

The *centroid of a volume* of a homogeneous material is the same as its center of gravity.

C
unitless

The *discharge coefficient* is a factor which relates the actual flow rate through a measuring device to the theoretically predicted flow rate.

μ
lb-sec/ft²

Dynamic viscosity is the ratio of the shearing stress in a fluid to the velocity gradient $\Delta v / \Delta y$.

z
ft

Elevation is the vertical distance from some reference level to a point of interest measured positive in the upward direction.

L_e
ft

Equivalent length is the length of straight pipe which would have the same total energy loss as a valve or fitting.

f
unitless

Friction factor is a dimensionless coefficient which indicates the resistance to the flow of a fluid. Energy losses due to friction are directly proportional to the friction factor.

N_F
unitless

The *Froude number* characterizes the state of open channel flow.

$$N_F = v/\sqrt{gy_h}$$

If $N_F = 1.0$ the flow is critical.
If $N_F < 1.0$ the flow is subcritical.
If $N_F > 1.0$ the flow is supercritical.

p_{gage}
psig

Gage pressure is measured relative to atmospheric pressure.

y_h
ft

Hydraulic depth is the ratio of the cross-sectional area of an open channel to the width of the free surface.

$$y_h = A/T$$

Hydraulic jump is an open channel flow phenomenon in which the flow stream abruptly changes from a smaller to a greater depth with a corresponding decrease in the velocity of flow and a dissipation of energy.

R
ft

Hydraulic radius is the ratio of the net cross-sectional area of a flow stream to the wetted perimeter of the section.

ν
ft²/sec

Kinematic viscosity is the ratio of the dynamic viscosity to the density of the fluid.

$$\nu = \mu/\rho$$

C_L
unitless

The *loss coefficient* is a constant of proportionality between a minor energy loss and the velocity head of the flowing fluid.

A *manometer* is a device which employs the height of a column of a static fluid to indicate the magnitude of a pressure.

M
slugs/sec

The *mass flow rate* is the mass of fluid flowing past a section per unit time.

mc	The *metacenter* is the intersection of the vertical axis of a floating body when in its equilibrium position and a vertical line through the new position of the center of buoyancy when the body is rotated slightly. It is a measure of the stability of the floating body.
	A *minor loss* is an energy loss due to a change in the cross section of the flow path or the direction of flow, or to an obstruction.
$\rho Q v$ lb	*Momentum flux* is the rate of transfer of momentum of a fluid.
y ft	*Normal depth* is the depth of an open channel flow stream at which uniform flow occurs for a particular discharge.
	Open channel flow occurs when a free surface of a flow stream is exposed to the atmosphere.
p lb/ft²	*Pressure* is defined by the equation $p = F/A$ where F is the force perpendicular to a surface and A is the area of the surface.
D/ϵ unitless	*Relative roughness* is the ratio of the diameter of a round pipe to the surface roughness of the pipe wall.
N_R unitless	*Reynolds number* is a dimensionless ratio which can be used to determine whether a flowing fluid is laminar or turbulent.

$$N_R = \frac{vD\rho}{\mu} = \frac{vD}{\nu}$$

If $N_R < 2000$ the flow is laminar.
If $N_R > 4000$ the flow is turbulent.

ϵ ft	The *roughness* of a surface is the average height of the peaks of the surface irregularities.
y ft	*Sequent depth* is the resulting depth of a flow stream after hydraulic jump occurs.

S
unitless

The *slope* of a channel is the ratio of the vertical drop to the length of the channel.

Steady flow occurs when the quantity of fluid flowing past any section in a given amount of time is constant.

T
ft

The *top width* is the width of the free surface of an open channel flow stream.

$v^2/2g$
ft

Velocity head represents the kinetic energy of the fluid per pound of fluid flowing.

Viscosity is the property of a fluid which offers resistance to the relative motion of fluid molecules.

Q
ft³/sec

The *volume flow rate* is the volume of fluid flowing past a section per unit time.

W
lb/sec

The *weight flow rate* is the weight of fluid flowing past a section per unit time.

WP
ft

Wetted perimeter is the sum of the length of the boundaries of a section actually in contact with (that is, wetted by) the fluid.

3

Fluids at Rest

Fluid Pressure

Pressure is defined by the equation

$$p = \frac{F}{A} \qquad\qquad \textbf{(3–1)}$$

where F is the force perpendicular to a surface and A is the area of the surface. Pressure in a fluid can be visualized by referring to Fig. 3–1 which shows a confined fluid supporting a load of 100 pounds acting on

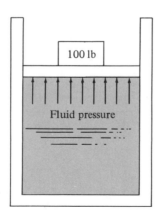

FIGURE 3–1

Illustration of Fluid Pressure

a piston which has an area of 5 square inches. All of the fluid in contact with the piston produces a reaction force opposing the applied force F. The pressure in the fluid at the piston is then,

$$p = \frac{F}{A} = \frac{100\ \text{lb}}{5\ \text{in.}^2} = 20\ \text{lb/in.}^2$$

The reaction force due to the fluid pressure acts perpendicular to the piston surface and therefore we can say that the pressure itself acts perpendicular to the surface.

In any fluid at rest the pressure at a boundary acts perpendicular to the boundary.

Figure 3–2 shows the direction of fluid pressure on the boundaries of several containers holding a confined fluid. These are common shapes and may represent the cross section of the following:

(a)   furnace duct
(b)   pipe or tube
(c)   heat exchanger (a pipe inside another pipe)
(d)   and (e)   swimming pool, storage tank, or reservoir
(f)   dam
(g)   fluid power cylinder

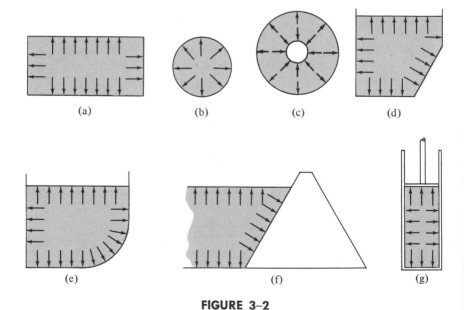

FIGURE 3–2

Direction of Fluid Pressure on Boundaries

Notice that in Fig. 3–2 (d), (e), and (f), pressure is shown acting at the surface of the liquid even though it is exposed to the atmosphere and not confined by a solid boundary. The interface between the liquid and the air is considered a boundary and the air exerts an equal and opposite pressure on the liquid. The sketches in Fig. 3–2 show only the direction of the fluid pressure. The magnitude varies within the fluid as will be shown in Section 3–3.

The units for pressure are always those of force per unit area such as pounds per square inch or pounds per square foot. Unfortunately, many people speak of so many "pounds of pressure" when they really mean pounds per square inch.

3–2   Absolute and Gage Pressure

When making calculations involving pressure in a fluid the measurements must be made relative to some reference pressure. Normally the reference pressure is that of the atmosphere and the resulting measured pressure is called *gage pressure*. Pressure measured relative to a perfect vaccum is called *absolute pressure*. It is extremely important for you to know the difference between these two ways of measuring pressure and to be able to convert from one to the other.

A simple equation relates the two pressure measuring systems:

$$p_{\text{abs}} = p_{\text{gage}} + p_{\text{atm}} \qquad\qquad (3\text{–}2)$$

where

$$p_{\text{abs}} = \text{absolute pressure}$$
$$p_{\text{gage}} = \text{gage pressure}$$
$$p_{\text{atm}} = \text{atmospheric pressure}$$

Figure 3–3 shows an interpretation of this equation graphically. A few basic concepts may help you to understand the equation.

1. A perfect vacuum is the lowest possible pressure. Therefore, an absolute pressure will always be positive.
2. A gage pressure above atmospheric pressure is positive.
3. A gage pressure below atmospheric pressure is negative, sometimes called vacuum.
4. Gage pressure is usually expressed in the units of pounds per square inch gage (psig). The units of inches of mercury vacuum are often used for pressures below atmospheric. (See Section 3–4.)
5. Absolute pressure is usually expressed in the units of pounds per square inch absolute (psia).

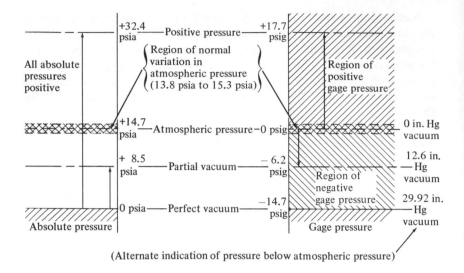

(Alternate indication of pressure below atmospheric pressure)

FIGURE 3–3

Absolute and Gage Pressure

6. The actual magnitude of the atmospheric pressure varies with location and with climatic conditions. The barometric pressure as broadcast in weather reports is an indication of the continually varying atmospheric pressure.
7. Unless the prevailing atmospheric pressure is given as data it will be assumed to be 14.7 psia in this book. The range of normal variation of atmospheric pressure near the earth's surface is approximately from 13.8 psia to 15.3 psia.

EXAMPLE PROBLEMS

Try these problems and then look at the solutions.

Example Problem 3–1. Express a pressure of 21.9 psig as an absolute pressure. The local atmospheric pressure is 14.5 psia.

Solution.

$$p_{abs} = p_{gage} + p_{atm}$$
$$p_{abs} = 21.9 \text{ psig} + 14.5 \text{ psia} = 36.4 \text{ psia}$$

Notice that the units in this calculation are pounds per square inch (lb/in.²) for each term and are consistent. The indication of gage or absolute is merely for convenience.

Example Problem 3–2. Express a pressure of 32.4 psia as a gage pressure. The local atmospheric pressure is 14.7 psia.

Solution.

$$p_{abs} = p_{gage} + p_{atm}$$

Solving algebraically for p_{gage} gives

$$p_{gage} = p_{abs} - p_{atm}$$
$$p_{gage} = 32.4 \text{ psia} - 14.7 \text{ psia} = 17.7 \text{ psig}$$

Example Problem 3–3. Express a pressure of 10.9 psia as a gage pressure. The local atmospheric pressure is 15.0 psia.

Solution.

$$p_{abs} = p_{gage} + p_{atm}$$
$$p_{gage} = p_{abs} - p_{atm}$$
$$p_{gage} = 10.9 \text{ psia} - 15.0 \text{ psia} = -4.1 \text{ psig}$$

Notice that the result is negative. This can also be read "4.1 psi below atmospheric pressure."

Example Problem 3–4. Express a pressure of $- 6.2$ psig as an absolute pressure.

Solution.

$$p_{abs} = p_{gage} + p_{atm}$$

Since no value was given for the atmospheric pressure we will use $p_{atm} = 14.7$ psia.

$$p_{abs} = -6.2 \text{ psig} + 14.7 \text{ psia} = 8.5 \text{ psia}$$

3–3   Relationship Between Pressure and Elevation

You are probably familiar with the fact that as you go deeper in a fluid, such as in a swimming pool, the pressure increases. There are many situations in which it is important to know just how pressure varies with a change in depth or elevation.

In this book the term elevation means the vertical distance from some reference level to a point of interest and is called z. A change in elevation between two points is called h. Elevation will always be measured positive in the upward direction. In other words, a higher point has a larger elevation than a lower point.

The reference level can be taken at any level as illustrated in Fig. 3–4 which shows a submarine under water. In part (a) of the figure the sea bottom is taken as reference while in (b) the position of the submarine is the reference level. Since fluid mechanics calculations usually consider differences in elevation, it is advisable to choose the lowest point of interest in a problem as the reference level to eliminate the use of negative values for z. This will be especially important in later work.

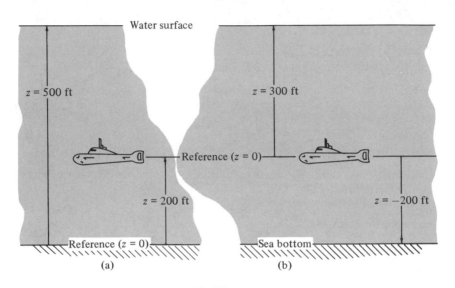

FIGURE 3–4

Illustration of Reference Level for Elevation

The change in pressure in a liquid at rest due to a change in elevation can be calculated from

$$\Delta p = \gamma h \qquad (3–3)$$

where

Δp = change in pressure
γ = specific weight of liquid
h = change in elevation

Some general conclusions from Eq. (3–3) will help you to apply it properly.

1. The equation is valid only for a homogeneous liquid at rest.
2. The change in pressure is directly proportional to the specific weight of the liquid.

3. Pressure varies linearly with the change in elevation.
4. A decrease in elevation causes an increase in pressure. (This is what happens when you go deeper in a swimming pool.)
5. An increase in elevation causes a decrease in pressure.
6. Points on the same horizontal level have the same pressure.

Equation (3–3) does not apply to gases because the specific weight of a gas changes with a change in pressure. However, it requires a large change in elevation to produce a significant change in pressure in a gas. For example, an increase in elevation of 1000 feet in the atmosphere causes a decrease in pressure of only 0.5 lb/in.² In this book it is assumed that the pressure in a gas is uniform unless otherwise specified.

EXAMPLE PROBLEMS

Example Problem 3–5. Calculate the change in water pressure from the surface to a depth of 15 feet.

Solution. Using Eq. (3–3), $\Delta p = \gamma h$, let $\gamma = 62.4$ lb/ft³ for water and $h = 15$ feet. Then

$$\Delta p = \frac{62.4 \text{ lb}}{\text{ft}^3} \cdot 15 \text{ ft} \cdot \frac{1 \text{ ft}^2}{144 \text{ in.}^2} = 6.5 \frac{\text{lb}}{\text{in.}^2}$$

If the surface of the water is exposed to the atmosphere, the pressure there is 0 psig. Going down in the water (decreasing elevation) produces an increase in pressure. Therefore, at 15 feet the pressure is 6.5 psig.

Example Problem 3–6. Figure 3–5 shows a tank of oil having one side open to the atmosphere and the other side sealed with air above the

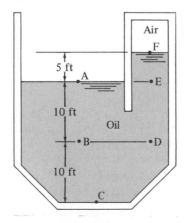

FIGURE 3–5

oil. The oil has a specific gravity of 0.90. Calculate the pressure in psig
at points A, B, C, D, E, and F and the air pressure in the right side of the
tank.

Solution. At point A the oil is exposed to the atmosphere and therefore
$p_A = 0$ psig.

Point B: The change in elevation between point A and point B is 10 feet
with B lower than A. To use Eq. (3–3) the specific weight of the oil is
needed.

$$\gamma_{oil} = (sg)_{oil}(62.4 \text{ lb/ft}^3)$$
$$\gamma_{oil} = (0.90)(62.4 \text{ lb/ft}^3) = 56.2 \text{ lb/ft}^3$$

Then,

$$\Delta p_{A-B} = \gamma h = \frac{56.2 \text{ lb}}{\text{ft}^3} \cdot \frac{10 \text{ ft}}{1} \cdot \frac{1 \text{ ft}^2}{144 \text{ in.}^2} = \frac{+3.9 \text{ lb}}{\text{in.}^2}$$

Now,

$$p_B = p_A + \Delta p_{A-B}$$
$$= 0 \text{ psig} + 3.9 \text{ lb/in.}^2$$
$$p_B = 3.9 \text{ psig}$$

Point C: The change in elevation from point A to point C is 20 feet with
C lower than A.
Then,

$$\Delta p_{A-C} = \gamma h = \frac{56.2 \text{ lb}}{\text{ft}^3} \cdot 20 \text{ ft} \cdot \frac{1 \text{ ft}^2}{144 \text{ in.}^2} = \frac{7.8 \text{ lb}}{\text{in.}^2}$$
$$p_C = p_A + \Delta p_{A-C}$$
$$= 0 \text{ psig} + 7.8 \text{ lb/in.}^2$$
$$p_C = 7.8 \text{ psig}$$

Point D: Since point D is at the same level as point B, the pressure is the
same. That is,

$$p_D = p_B = 3.9 \text{ psig}$$

Point E: Since point E is at the same level as point A, the pressure is the
same. That is,

$$p_E = p_A = 0 \text{ psig}$$

Point F: The change in elevation between point A and point F is 5 feet
with F higher than A. Then,

$$\Delta p_{A-F} = -\gamma h = \frac{-56.2 \text{ lb}}{\text{ft}^3} \cdot 5 \text{ ft} \cdot \frac{1 \text{ ft}^2}{144 \text{ in.}^2} = -1.95 \text{ lb/in.}^2$$
$$p_F = p_A + \Delta p_{A-F} = 0 \text{ psig} + (-1.95) \text{ lb/in.}^2$$
$$p_F = -1.95 \text{ psig}$$

Air pressure: Since the air in the right side of the tank is exposed to the surface of the oil where $p_F = -1.95$ psig, the air pressure is also -1.95 psig or 1.95 psi below atmospheric pressure.

3–4   Manometers, Barometers, and Pressure Gages

There are many types of devices for measuring pressure. Several commonly used types will be described here.

Manometers. The relationship between a change in pressure and a change in elevation as described in Section 3–3 is the principle on which a manometer is based. The simplest kind is called a U-tube manometer as shown in Fig. 3–6. One end of the U-tube is connected to the pressure which is to be measured while the other end is left open to the atmosphere. The tube contains a liquid called the gage fluid which does not mix with the fluid whose pressure is to be measured. Typical gage fluids are water, mercury, and colored light oils.

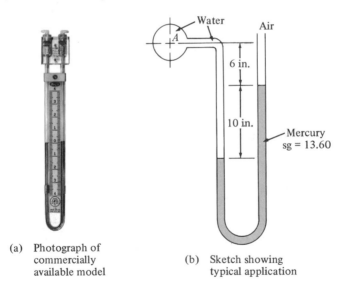

(a) Photograph of
     commercially
     available model

(b) Sketch showing
     typical application

FIGURE 3–6

U-Tube Manometer

Source of photo: F. W. Dwyer Mfg. Co.
Michigan City, Indiana

Under the action of the pressure to be measured, the gage fluid is displaced from its normal position. Since the fluids in the manometer are at rest, the equation $\Delta p = \gamma h$ can be used to write expressions for the changes in pressure that occur throughout the manometer. These expressions can then be combined and solved algebraically for the desired pressure. This general approach can be set up in a step-by-step procedure which you should learn since manometers are used in many real situations such as those described in this book.

1. Start from a convenient point, usually where the pressure is known, and write this pressure in symbol form. (e.g., p_A refers to the pressure at point A.)
2. Using $\Delta p = \gamma h$, write expressions for the changes in pressure that occur from the starting point to the point at which the pressure is to be measured, being careful to include the correct algebraic sign for each term.
3. Equate the expression from Step 2 to the pressure at the desired point.
4. Substitute known values and solve for the desired pressure.

The working of several practice problems will help you to apply this procedure correctly. The following problems are written in the programmed instruction format. To work through the program, cover the material below the heading "Programmed Example Problems" and then proceed one panel at a time.

Programmed Example Problems

Example Problem 3–7. Using Fig. 3–6 calculate the pressure at point A in psig.

Perform Step 1 of the procedure before going to the next panel.

Figure 3–7 is identical to Fig. 3–6(b) except that certain key points have been numbered for use in the problem solution.

The only point for which the pressure is known is the surface of the mercury in the right leg of the manometer, point 1. Now, how can an expression be written for the pressure that exists within the mercury, 10 inches below this surface at point 2?

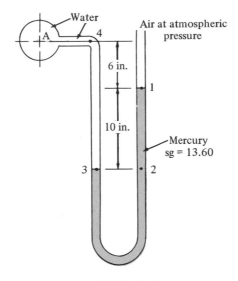

Water

A

4

Air at atmospheric
pressure

6 in.

1

10 in.

Mercury
sg = 13.60

3

2

FIGURE 3–7

U-Tube Manometer

The expression is

$$p_1 + \gamma_m(10 \text{ in.})$$

The term γ_m (10 in.) is the change in pressure between points 1 and 2 due to a change in elevation, where γ_m is the specific weight of mercury, the gage fluid. This pressure change is added to p_1 because there is an increase in pressure as we go down in a fluid.

So far we have an expression for the pressure at point 2 in the right leg of the manometer. Now write the expression for the pressure at point 3 in the left leg.

This is the expression:

$$p_1 + \gamma_m(10 \text{ in.})$$

Because points 2 and 3 are on the same level in the same fluid at rest, their pressures are equal.

Continue on and write the expression for the pressure at point 4.

$$p_1 + \gamma_m(10 \text{ in.}) - \gamma_w(16 \text{ in.})$$

where γ_w is the specific weight of water. Remember, there is a decrease in pressure between points 3 and 4 so this last term must be subtracted from our previous expression.

What must be done to get an expression for the pressure at A?

Nothing. Since points A and 4 are on the same level their pressures are equal.

Now perform Step 3 of the procedure.

You should now have

$$p_1 + \gamma_m(10 \text{ in.}) - \gamma_w(16 \text{ in.}) = p_A$$

or,

$$p_A = p_1 + \gamma_m(10 \text{ in.}) - \gamma_w(16 \text{ in.})$$

This is the completed equation for the pressure at A. Now do Step 4.

Several calculations are required here.

$$p_1 = p_{\text{atm}} = 0 \text{ psig}$$
$$\gamma_m = (sg)_m(62.4 \text{ lb/ft}^3) = (13.60)(62.4 \text{ lb/ft}^3)$$
$$= 849 \text{ lb/ft}^3$$
$$\gamma_w = 62.4 \text{ lb/ft}^3$$

Then,

$$p_A = p_1 + \gamma_m(10 \text{ in.}) - \gamma_w(16 \text{ in.})$$
$$p_A = 0 \text{ psig} + \frac{849 \text{ lb}}{\text{ft}^3} \cdot 10 \text{ in.} \cdot \frac{1 \text{ ft}^3}{1728 \text{ in.}^3} - \frac{62.4 \text{ lb}}{\text{ft}^3}$$
$$\cdot 16 \text{ in.} \cdot \frac{1 \text{ ft}^3}{1728 \text{ in.}^3}$$
$$= 0 \text{ psig} + 4.91 \frac{\text{lb}}{\text{in.}^2} - 0.58 \frac{\text{lb}}{\text{in.}^2}$$
$$p_A = 4.33 \text{ lb/in.}^2 = 4.33 \text{ psig.}$$

Remember to include the units in your calculations. Review this problem to be sure of understanding every step before going to the next panel for another problem.

Example Problem 3–8. Calculate the pressure at point B in Fig. 3–8 if the pressure at A is 22.40 psig. This type of manometer is called a differential manometer because it indicates the difference between the pressure at A and B but not the actual value of either one.

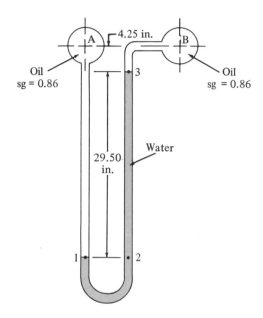

FIGURE 3–8

Differential Manometer

Do Step 1 of the procedure to find p_B.

We know the pressure at A so we start there and call it p_A. Now write an expression for the pressure at point 1.

You should have

$$p_A + \gamma_o \ (33.75 \ \text{in.})$$

where γ_o is the specific weight of the oil.
What is the pressure at point 2?

It is the same as the pressure at point 1 because the two points are on the same level. Go on to point 3 in the manometer.

The expression should now look like this.

$$p_A + \gamma_o \, (33.75 \text{ in.}) - \gamma_w \, (29.5 \text{ in.})$$

Now write the expression for the pressure at point 4.

This is the desired expression.

$$p_A + \gamma_o \, (33.75 \text{ in.}) - \gamma_w \, (29.5 \text{ in.}) - \gamma_o \, (4.25 \text{ in.})$$

This is also the expression for the pressure at B since points 4 and B are on the same level. Now do Steps 3 and 4 of the procedure.

Our final expression should be

$$p_A + \gamma_o(33.75 \text{ in.}) - \gamma_w(29.5 \text{ in.}) - \gamma_o(4.25 \text{ in.}) = p_B$$

or,

$$p_B = p_A + \gamma_o(33.75 \text{ in.}) - \gamma_w(29.5 \text{ in.}) - \gamma_o(4.25 \text{ in.})$$

The known values are:

$p_A = 22.40$ psig
$\gamma_o = (sg)_o(62.4 \text{ lb/ft}^3) = (0.86)(62.4 \text{ lb/ft}^3) = 53.6 \text{ lb/ft}^3$
$\gamma_w = 62.4 \text{ lb/ft}^3$

In this case it may help to simplify the expression for p_B before substituting known values. Since two terms are multiplied by γ_o they can be combined.

$$p_B = p_A + \gamma_o(29.5 \text{ in.}) - \gamma_w(29.5 \text{ in.})$$

Now factoring out the 29.5 in. term gives

$$p_B = p_A + (29.5 \text{ in.})(\gamma_o - \gamma_w)$$

This looks simpler than the original equation. The difference between p_A and p_B is a function of the *difference* between the specific weights of the two fluids.

Now,

$$p_B = 22.40 \text{ psig} + (29.5 \text{ in.})(53.6 - 62.4) \frac{\text{lb}}{\text{ft}^3} \cdot \frac{1 \text{ ft}^3}{1728 \text{ in.}^3}$$

$$= 22.40 \text{ psig} + \frac{(29.5)(-8.8) \text{ lb/in.}^2}{1728}$$

$$= 22.40 \text{ psig} - 0.15 \text{ lb/in.}^2$$

$$p_B = 22.25 \text{ psig}$$

Notice that using a gage fluid having a specific weight very close to that of the fluid being measured makes the manometer very sensitive. This concludes the programmed instruction.

Figure 3–9 shows another type of manometer called the well-type. When a pressure is applied to a well-type manometer, the fluid level in the well drops a small amount while the level in the right leg rises a larger amount in proportion to the ratio of the areas of the well and the tube. A scale is placed alongside the tube so that the deflection can be read directly. The scale is calibrated to account for the small drop in the well level.

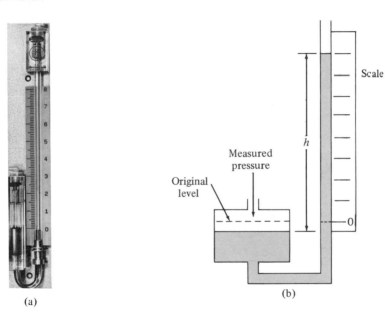

(a)

(b)

FIGURE 3–9

Well-Type Monometer

Source of photo: F. W. Dwyer Mfg. Co.
                   Michigan City, Ind.

The inclined well-type manometer, Fig. 3–10, has the same features as the well-type but offers a greater sensitivity by placing the scale along the inclined tube. The scale length is increased as a function of the

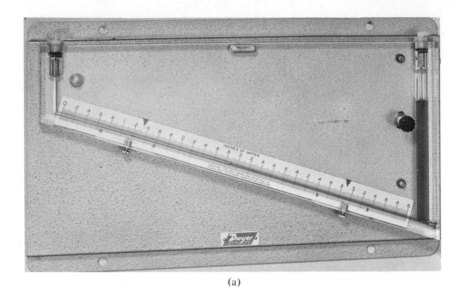

(a)

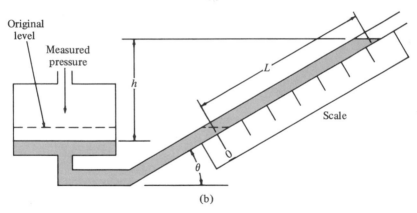

(b)

FIGURE 3–10

Inclined Well-Type Manometer

Source of photo: F. W. Dwyer Mfg. Co.
            Michigan City, Ind.

angle of inclination of the tube, θ. For example, if the angle θ in Fig.
3–10(b) is 15° the ratio of scale length L to manometer deflection h is

$$\frac{h}{L} = \sin \theta$$

or

$$\frac{L}{h} = \frac{1}{\sin \theta} = \frac{1}{\sin 15°} = \frac{1}{0.259} = 3.86$$

The scale would be calibrated so that the deflection could be read directly.

Barometers. A device for measuring the atmospheric pressure is called a barometer. A simple type is shown in Fig. 3–11. It consists of a long tube closed at one end which is initially filled completely with mercury.

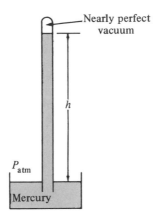

FIGURE 3–11

Barometer

The open end is then submerged under the surface of a container of mercury and allowed to come to equilibrium as shown in Fig. 3–11. A void is produced at the top of the tube which is very nearly a perfect vacuum, containing mercury vapor at a pressure of only 0.000025 psia at 70°F. By starting at this point and writing an equation similar to those for manometers, we get

$$0 + \gamma_m h = p_{atm}$$

or,

$$p_{atm} = \gamma_m h \tag{3–4}$$

Since the specific weight of mercury is approximately constant, a change in atmospheric pressure will cause a change in the height of the mercury column. This height is often reported as the barometric pressure. To obtain true atmospheric pressure it is necessary to multiply h by γ_m.

EXAMPLE PROBLEMS

Example Problem 3–9. A news broadcast reported that the barometric pressure is 30.40 inches of mercury. Calculate the atmospheric pressure in psia.

Solution.   In Eq. (3–4),

$$p_{atm} = \gamma_m h,$$
$$\gamma_m = 849 \text{ lb/ft}^3 \text{ and } h = 30.40 \text{ in.}$$

Then,

$$p_{atm} = \frac{849 \text{ lb}}{ft^3} \cdot 30.40 \text{ in.} \cdot \frac{1 \text{ ft}^3}{1728 \text{ in.}^3} = 14.9 \text{ lb/in.}^2$$
$$p_{atm} = 14.9 \text{ psia.}$$

Example Problem 3–10.   The standard atmospheric pressure is 14.7 psia. Calculate the height of a mercury column equivalent to this pressure.

Solution.

$$p_{atm} = \gamma_m h$$
$$h = \frac{p_{atm}}{\gamma_m} = \frac{14.7 \text{ lb}}{\text{in.}^2} \cdot \frac{ft^3}{849 \text{ lb}} \cdot \frac{1728 \text{ in.}^3}{ft^3}$$
$$h = 29.92 \text{ in.}$$

It may be well to remember that 29.92 inches of mercury is equivalent to 14.7 psia. It can be used as a conversion factor as illustrated in the next problem.

Example Problem 3–11.   Convert a barometric pressure of 28.5 inches of mercury to a pressure in psia.

$$28.50 \text{ in. of mercury} \cdot \frac{14.7 \text{ psia}}{29.92 \text{ in. of mercury}} = 14.0 \text{ psia}$$

Pressure Gages.   Probably the most widely used pressure measuring device is the Bourdon tube pressure gage as shown in Fig. 3–12. The pressure to be measured is applied to the inside of a flattened tube which is normally shaped as a segment of a circle or a spiral. The increased pressure inside the tube causes it to be straightened somewhat. The movement of the end of the tube is transmitted through a linkage which causes a pointer to rotate.

The scale of the gage normally reads zero when the gage is open to atmospheric pressure and is calibrated in pounds per square inch pressure above zero. Therefore, this type of gage reads *gage pressure* directly. Some gages are capable of reading pressures below atmospheric. The scale below zero is usually calibrated in "inches of mercury vacuum." The range is from zero down to 30 inches of mercury which corresponds approximately to a perfect vacuum. However, the readings are relative to the prevailing atmospheric pressure which varies. In most cases the accuracy required in a measurement does not warrant correcting for the changing atmospheric pressure.

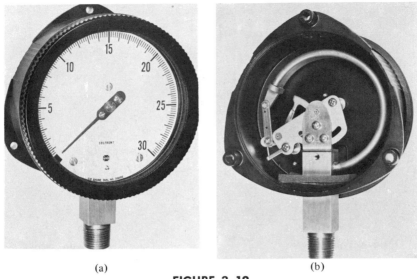

(a) (b)

FIGURE 3–12

Bourdon Tube Pressure Gage

Source of photo: Ametek/U.S. Gauge
Sellersville, Pa.

3–5   Forces on Submerged Plane Areas Due to Fluid Pressure

Pressure has been defined by the equation $p = F/A$ where F is a force, and A is the area on which the force acts. Figure 3–13 illustrates several situations in which it is necessary to calculate the forces that are exerted on a plane (flat) area due to fluid pressure. The magnitude and distribution of these forces must be known in order to properly design the components of each item.

In this section three important special cases of forces on submerged plane areas will be discussed before a general procedure is presented.

Gases Under Pressure.   Figure 3–14 shown a pneumatic cylinder of the type used in automated machinery. The air pressure acts on the piston face producing a force which causes the linear movement of the rod. The pressure also acts on the end of the cylinder tending to pull it apart. This is the reason for the four tie rods between the end caps of the cylinder. The distribution of pressure within a gas is very nearly uniform. Therefore, we can calculate the force on the piston and the cylinder ends directly from $F = pA$.

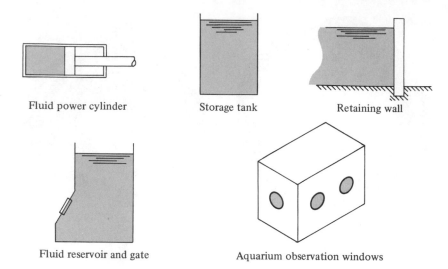

Fluid power cylinder   Storage tank   Retaining wall

Fluid reservoir and gate   Aquarium observation windows

FIGURE 3–13

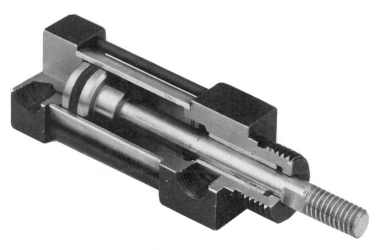

FIGURE 3–14

Fluid Power Cylinder

Source of photo: Mosier Industries
Dayton, Ohio

Example Problem 3–12. If the cylinder in Fig. 3–14 has an internal
diameter of 2 inches and operates at a pressure of 300 psig, calculate
the force on the ends of the cylinder.

Solution.

$$F = pA$$

$$A = \frac{\pi D^2}{4} = \frac{\pi(2 \text{ in.})^2}{4} = 3.14 \text{ in.}^2$$

$$F = \frac{300 \text{ lb}}{\text{in.}^2} \cdot 3.14 \text{ in.}^2 = 944 \text{ lb}$$

Notice that gage pressure was used in the calculation of force instead of absolute pressure. The additional force due to atmospheric pressure acts on both sides of the area and is thus balanced. If the pressure on the outside surface is not atmospheric, then all external forces must be considered to determine a net force on the area.

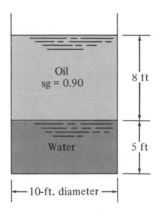

FIGURE 3–15

Horizontal Flat Surfaces Under Liquids. Figure 3–15 shows a cylindrical drum containing oil and water. The pressure in the water at the bottom of the drum is uniform across the entire area since it is a horizontal plane in a fluid at rest. Again we can simply use $F = pA$ to calculate the force on the bottom.

Example Problem 3–13. If the drum in Fig. 3–15 is open to the atmosphere at the top, calculate the force on the bottom.

Solution. To use $F = pA$ we must first calculate the pressure at the bottom of the drum p_B, and the area of the bottom.

$$p_B = p_{\text{atm}} + \gamma_o(8 \text{ ft}) + \gamma_w(5 \text{ ft})$$
$$\gamma_o = (\text{sg})_o(62.4 \text{ lb/ft}^3) = (0.90)(62.4 \text{ lb/ft}^3) = 56.2 \text{ lb/ft}^3$$
$$p_B = \frac{0 \text{ lb}}{\text{ft}^2} (\text{gage}) + \frac{56.2 \text{ lb}}{\text{ft}^3} \cdot 8 \text{ ft} + \frac{62.4 \text{ lb}}{\text{ft}^3} \cdot 5 \text{ ft}$$

$$p_B = (0 + 450 + 312) \text{ lb/ft}^2 = 762 \text{ lb/ft}^2$$
$$A = \pi D^2/4 = \pi(10 \text{ ft})^2/4 = 78.5 \text{ ft}^2$$
$$F = p_B A = \frac{762 \text{ lb}}{\text{ft}^2} \cdot 78.5 \text{ ft}^2 = 59,800 \text{ lb}$$

Example Problem 3–14. Would there be any difference between the force on the bottom of the drum in Fig. 3–15 and that on the bottom of the cone-shaped container in Fig. 3–16?

Solution. The force would be the same since the pressure at the bottom is dependent only upon the depth and specific weight of the fluid in the container. The total weight of fluid has no effect.

Rectangular Walls. The retaining walls shown in Fig. 3–17 are typical examples of rectangular walls which are exposed to a pressure varying

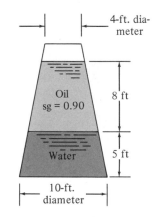

FIGURE 3–16

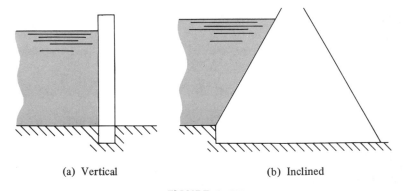

(a) Vertical   (b) Inclined

FIGURE 3–17

Rectangular Walls

from zero at the surface of the fluid to a maximum at the bottom of the wall. The force due to the fluid pressure tends to overturn the wall or break it at the place where it is fixed to the bottom.

The actual force is distributed over the entire wall, but for the purpose of analysis it is desirable to determine the resultant force and the place where it acts, called the center of pressure. That is, if the entire force was concentrated at a single point, where would that point be and what would the magnitude of the force be?

Figure 3–18 shows the pressure distribution on the vertical retaining wall. As indicated by the equation $\Delta p = \gamma h$, the pressure varies linearly (in a straight line manner) with depth in the fluid. The lengths of the

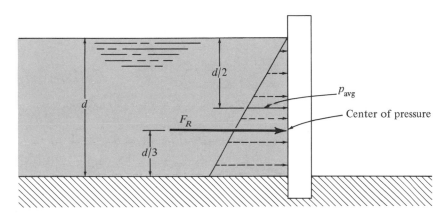

FIGURE 3–18

Vertical Rectangular Wall

dashed arrows represent the magnitude of the fluid pressure at various points on the wall. Because of this linear variation in pressure, the total resultant force can be calculated by the equation,

$$F_R = p_{\text{avg}} \cdot A \qquad (3\text{–}5)$$

where p_{avg} is the average pressure and A is the total area of the wall. But the average pressure is that at the middle of the wall and can be calculated by the equation,

$$p_{\text{avg}} = \gamma(d/2) \qquad (3\text{–}6)$$

where d is the total depth of the fluid.

Therefore,

$$F_R = \gamma(d/2)A \qquad (3\text{–}7)$$

The pressure distribution shown in Fig. 3–18 indicates that a greater portion of the force acts on the lower part of the wall than on the upper part. The center of pressure is at the centroid of the pressure distribution triangle, one-third of the distance from the bottom of the wall. The resultant force F_R acts perpendicular to the wall at this point.

The procedure for calculating the magnitude of the resultant force due to fluid pressure and the location of the center of pressure on a rectangular wall such as those shown in Fig. 3–17 is listed below. The procedure applies whether the wall is vertical or inclined.

1. Calculate the magnitude of the resultant force F_R from

$$F_R = \gamma(d/2)A$$

   where

$$\gamma = \text{specific weight of the fluid}$$
$$d = \text{total depth of the fluid}$$
$$A = \text{total area of the wall}$$

2. Locate the center of pressure at a distance of $d/3$ from the bottom of a vertical wall, or $\ell/3$ from the bottom of an inclined wall where ℓ is the length of the face of the wall.
3. Show the resultant force acting at the center of pressure perpendicular to the wall.

Example Problem 3–15.   In Fig. 3–18, the fluid is gasoline (sg = 0.68) and the total depth is 12 feet. The wall is 40 feet long. Calculate the magnitude of the resultant force on the wall and the location of the center of pressure.

Solution.

Step 1.

$$F_R = \gamma(d/2)A$$
$$\gamma = (0.68)(62.4 \text{ lb/ft}^3) = 42.4 \text{ lb/ft}^3$$
$$A = (12 \text{ ft})(40 \text{ ft}) = 480 \text{ ft}^2$$
$$F_R = \frac{42.4 \text{ lb}}{\text{ft}^3} \cdot \frac{12 \text{ ft}}{2} \cdot 480 \text{ ft}^2 = 122,000 \text{ lb}$$

Step 2.

The center of pressure is at a distance of

$$d/3 = 12 \text{ ft}/3 = 4 \text{ ft}$$

from the bottom of the wall.

Step 3.

The force F_R acts perpendicular to the wall at the center of pressure as shown in Fig. 3–18.

Example Problem 3–16. Figure 3–19 shows a dam, 100 feet long, which retains 27 feet of fresh water and is inclined at an angle, θ, of 60°. Calculate the magnitude of the resultant force on the dam and the location of the center of pressure.

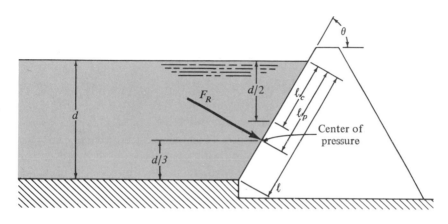

FIGURE 3–19

Inclined Rectangular Wall

Solution.

Step 1.

$$F_R = \gamma(d/2)A$$

To calculate the area of the dam we need the height of its face, called ℓ in Fig. 3–19.

$$\sin \theta = d/\ell$$
$$\ell = d/\sin \theta = 27 \text{ ft}/\sin 60° = 31.2 \text{ ft}$$

Then,

$$A = (31.2 \text{ ft})(100 \text{ ft}) = 3120 \text{ ft}^2$$

Now,

$$F_R = \gamma(d/2)A = \frac{62.4 \text{ lb}}{\text{ft}^3} \cdot \frac{27 \text{ ft}}{2} \cdot 3120 \text{ ft}^2$$
$$F_R = 2,630,000 \text{ lb}$$

Step 2.

The center of pressure is at a vertical distance of

$$d/3 = 27 \text{ ft}/3 = 9 \text{ ft}$$

from the bottom of the dam.

Measured along the face of the dam, the following dimensions are used as shown in Fig. 3–19.

ℓ_c = distance from free surface of the fluid to the center of the wall

$\ell_c = \ell/2 = 31.2 \text{ ft}/2 = 15.6 \text{ ft}$

ℓ_p = distance from the free surface of the fluid to the center of pressure

$\ell_p = \ell - d/3 \sin \theta = 31.2 \text{ ft} - 27 \text{ ft}/3 \sin 60°$

$\ell_p = 31.2 \text{ ft} - 10.4 \text{ ft} = 20.8 \text{ ft}$

Submerged Plane Areas — General. The procedure discussed in this section applies to problems dealing with plane areas, either vertical or inclined, which are completely submerged in the fluid. As in previous problems, the procedure will enable us to calculate the magnitude of

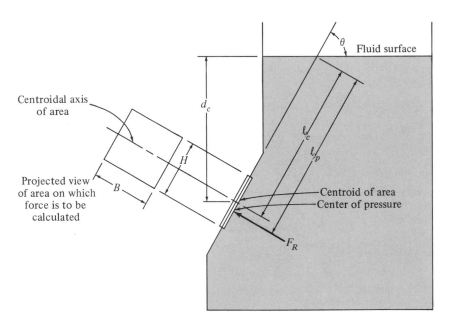

FIGURE 3–20

Force on Plane Submerged Area

the resultant force on an area and the location of the center of pressure where the resultant force can be assumed to act.

Figure 3–20 shows a tank which has a rectangular window in an inclined wall. The standard dimensions and symbols used in the procedure described later are shown in the figure and defined below.

F_R	Resultant force on the area due to the fluid pressure
—	The *center of pressure* of the area is the point at which the resultant force can be considered to act.
—	The *centroid* of the area is the point at which the area would be balanced if suspended from that point. It is equivalent to the center of gravity of a solid body.
θ	Angle of inclination of the area
d_c	Depth of fluid from the free surface to the centroid of the area
ℓ_c	Distance from the level of the free surface of the fluid to the centroid of the area, measured along the angle of inclination of the area.
ℓ_p	Distance from the level of the free surface of the fluid to the center of pressure of the area, measured along the angle of inclination of the area.
B,H	Dimensions of the area

Figure 3–21 shows the location of the centroid of several common geometrical shapes. Other shapes, and the method for calculating the location of the centroid of composite shapes, are described in the appendix.

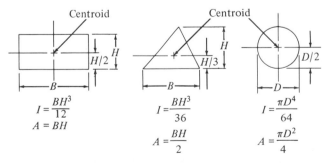

FIGURE 3–21

Properties of Simple Areas

The following is the procedure for calculating the magnitude of the resultant force on a submerged plane area due to fluid pressure and the location of the center of pressure.

1. Determine the centroid of the area.
2. Determine d_c and ℓ_c from the geometry of the problem.
3. Calculate the total area A on which the force is to be determined.
4. Calculate the resultant force from

$$F_R = \gamma d_c A \qquad (3\text{–}8)$$

where γ is the specific weight of the fluid. This equation states that the resultant force is the product of the pressure at the centroid of the area and the total area.

5. Calculate I, the moment of inertia of the area about its centroidal axis.
6. Calculate the location of the center of pressure from

$$\ell_p = \ell_c + \frac{I}{\ell_c A} \qquad (3\text{–}9)$$

Notice that the center of pressure is always below the centroid of an area which is inclined with the horizontal. In some cases it may be of interest to calculate only the difference between ℓ_p and ℓ_c from

$$\ell_p - \ell_c = \frac{I}{\ell_c A} \qquad (3\text{–}10)$$

7. Show the resultant force F_R on a sketch acting at the center of pressure, perpendicular to the area.
8. Show the dimension ℓ_p on the sketch in a manner similar to that used in Fig. 3–20.

The programmed instruction approach will now be used to illustrate the application of the procedure.

Programmed Example Problems

Example Problem 3–17. The tank shown in Fig. 3–20 contains a lubricating oil having a specific gravity of 0.91. The rectangular gate with the dimensions $B = 4$ ft and $H = 2$ ft is placed in the inclined wall of the tank ($\theta = 60°$). The centroid of the gate is at a depth of 5 feet from the surface of the oil. Calculate the magnitude of the resultant force F_R on the gate and the location of the center of pressure.

Using the procedure described above, perform Step 1 before going to the next panel.

The area of interest is the rectangular gate sketched as Fig. 3–22. The centroid is at the intersection of the axes of symmetry of the rectangle.

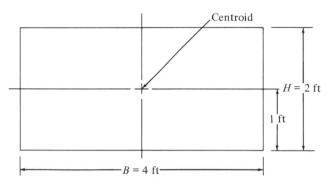

FIGURE 3–22

Now from Step 2, what is the distance d_c?

From the problem statement we know that $d_c = 5$ ft, the vertical depth from the free surface of the oil to the centroid of the gate.

Now calculate ℓ_c.

The terms ℓ_c and d_c are related in this case by

$$\sin \theta = d_c/\ell_c$$

Therefore,

$$\ell_c = d_c/\sin \theta = 5 \text{ ft}/\sin 60° = 5.77 \text{ ft}$$

Both d_c and ℓ_c will be needed for later calculations.

Go on to Step 3.

Since the area of the rectangle is BH,

$$A = BH = (4 \text{ ft})(2 \text{ ft}) = 8 \text{ ft}^2$$

Now do Step 4.

In the equation $F_R = \gamma d_c A$ the specific weight of the oil is needed.

$$\gamma_o = (sg)_o(62.4 \text{ lb/ft}^3) = (0.91)(62.4 \text{ lb/ft}^3)$$
$$\gamma_o = 56.8 \text{ lb/ft}^3$$

Then,

$$F_R = \gamma_o d_c A = \frac{56.8 \text{ lb}}{\text{ft}^3} \cdot 5 \text{ ft} \cdot 8 \text{ ft}^2 = 2270 \text{ lb}$$

The next steps concern the location of the center of pressure. Go on to Step 5 now.

From Fig. 3–21 we find that for a rectangle,

$$I = BH^3/12 = (4 \text{ ft})(2 \text{ ft})^3/12 = 2.67 \text{ ft}^4$$

Now we have all the data necessary to do Step 6.

The result is $\ell_p = 5.828$ ft.
Since $I = 2.67 \text{ ft}^4$, and $\ell_c = 5.77$ ft, and $A = 8 \text{ ft}^2$,

$$\ell_p = \ell_c + \frac{I}{\ell_c A} = 5.77 \text{ ft} + \frac{2.67 \text{ ft}^4}{(5.77 \text{ ft})(8 \text{ ft}^2)}$$
$$\ell_p = 5.77 \text{ ft} + 0.058 \text{ ft} = 5.828 \text{ ft}$$

This means that the center of pressure is 0.058 foot (0.70 inch) below the centroid of the gate.

Steps 7 and 8 are already completed in Fig. 3–20. Be sure you understand how the dimension ℓ_p is indicated.

The following problem is more complex in that the area on which the force is to be calculated is a composite area, rather than a simple geometric shape.

Example Problem 3–18. Figure 3–23 shows a water reservoir with a trapezoidal shaped gate in an inclined side. Calculate the magnitude of the resultant force on the gate and the location of the center of pressure.

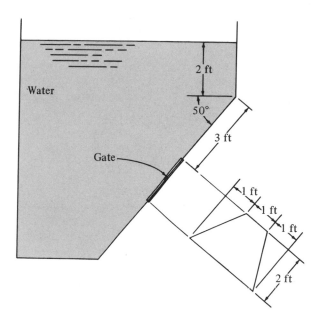

FIGURE 3–23

Perform Step 1 of the procedure before going to the next panel.

The correct location of the centroid of the gate is 0.833 feet from its bottom as indicated by the dimension $\bar{Y}$ in Fig. 3–24. There are several

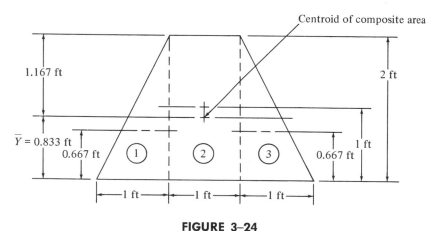

FIGURE 3–24

Gate Geometry

specific procedures for calculating this dimension. That detailed below
is done according to the method presented in the appendix.

The composite area is divided into three parts labeled 1, 2, and 3
in Fig. 3–24. The location of the centroid for each part is determined
and indicated by the dimension $\bar{y}$ from the bottom of the gate. The area
of each part is then calculated. This data is entered in a table as shown
below.

TABLE 3–1

Part	A	$\bar{y}$	$A\bar{y}$
1	1 ft²	0.667 ft	0.667 ft³
2	1 ft²	0.667 ft	0.667 ft³
3	2 ft²	1.000 ft	2.000 ft³
$\Sigma A = 4$ ft²			3.334 ft³ = $\Sigma(A\bar{y})$

Then,

$$\bar{Y} = \frac{\Sigma(A\bar{y})}{\Sigma A} = \frac{3.334 \text{ ft}^3}{4 \text{ ft}^2} = 0.833 \text{ ft}$$

Now do Step 2.

Figure 3–25 shows the dimensions d_c and ℓ_c which have the values
5.192 feet and 6.778 feet respectively. Again there are several ways of
determining these values, one of which is described below.

From Fig. 3–24 the distance from the top of the gate to its centroid
is 1.167 feet. The dimension ℓ_c is then

$$\ell_c = 1.167 \text{ ft} + 3.0 \text{ ft} + x$$

as shown in Fig. 3–25. But x is found from

$$\sin 50° = 2 \text{ ft}/x$$
$$x = 2 \text{ ft}/\sin 50° = 2.611 \text{ ft}$$

Then,

$$\ell_c = (1.167 + 3.0 + 2.611) \text{ ft} = 6.778 \text{ ft}$$

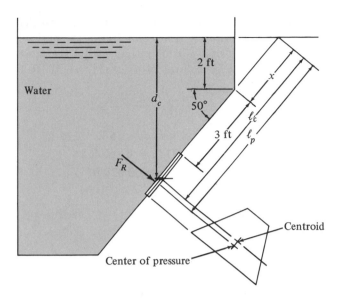

FIGURE 3–25

The dimensions d_c and ℓ_c are related by

$$\sin 50° = d_c/\ell_c$$

Then,

$$d_c = \ell_c \sin 50° = (6.778 \text{ ft})(\sin 50°) = 5.192 \text{ ft}$$

Now do Step 3 of the procedure.

Did you get $A = 4$ ft²? This was already done as a part of Step 1 above. We can now go on to Step 4.

Using $\gamma = 62.4$ lb/ft³ for water, the resultant force is

$$F_R = \gamma d_c A = \frac{62.4 \text{ lb}}{\text{ft}^3} \cdot 5.192 \text{ ft} \cdot 4.0 \text{ ft}^2 = 1296 \text{ lb}$$

Now begin to calculate the location of the center of pressure by completing Step 5.

The correct value for the total moment of inertia of the gate area about its centroidal axis is $I = 1.223$ ft⁴. This is found by addition of the moment of inertia values for each part, using the transfer theorem $I = I_c + Ay^2$, where I_c is the moment of inertia of the part with respect to its own centroidal axis, and y is the distance from the centroidal axis of the composite area to the centroidal axis of the part. Refer to Fig. 3–24 for dimensions.

Part 1:

$$I_c = BH^3/36 = (1 \text{ ft})(2 \text{ ft})^3/36 = 0.222 \text{ ft}^4$$
$$y = (0.833 - 0.667) \text{ ft} = 0.166 \text{ ft}$$
$$Ay^2 = (1.0 \text{ ft}^2)(0.166 \text{ ft})^2 = 0.028 \text{ ft}^4$$
$$I = (0.222 + 0.028) \text{ ft}^4 = 0.250 \text{ ft}^4$$

Part 2 (Same as Part 1):

$$I = 0.250 \text{ ft}^4$$

Part 3:

$$I_c = BH^3/12 = (1 \text{ ft})(2 \text{ ft})^3/12 = 0.667 \text{ ft}^4$$
$$y = (1.000 - 0.833) \text{ ft} = 0.167 \text{ ft}$$
$$Ay^2 = (2.0 \text{ ft}^2)(0.167 \text{ ft})^2 = 0.056 \text{ ft}^4$$
$$I = (0.667 + 0.056) \text{ ft}^4 = 0.723 \text{ ft}^4$$

Total for composite area:

$$I = (0.250 + 0.250 + 0.723) \text{ ft}^4 = 1.223 \text{ ft}^4$$

We can now find ℓ_p from Step 6.

$$\ell_p = \ell_c + \frac{I}{\ell_c A}$$
$$= 6.778 \text{ ft} + \frac{1.223 \text{ ft}^4}{(6.778 \text{ ft})(4.0 \text{ ft}^2)}$$
$$\ell_p = 6.778 \text{ ft} + 0.045 \text{ ft} = 6.823 \text{ ft}$$

This means that the center of pressure is 0.045 feet (0.54 inches) below the centroid of the gate.

Steps 7 and 8 are shown completed in Fig. 3–25. Be sure to notice that F_R is drawn perpendicular to the gate and that the distance ℓ_p is measured along the angle of inclination of the gate from the level of the free surface of the fluid.

This completes the programmed instruction.

3–6   Buoyancy

Buoyancy is the tendency for a fluid to exert a supporting force on a body placed in the fluid. This force is called a buoyant force and is defined by Archimedes' Principle as stated below.

A body in a fluid, whether floating or submerged, is buoyed up by a force equal to the weight of the fluid displaced. The buoyant force acts vertically upward through the centroid of the displaced volume and can be defined mathematically by the equation:

$$F_b = \gamma_f V_d \tag{3–11}$$

where

$$\gamma_f = \text{specific weight of the fluid}$$
$$V_d = \text{displaced volume of the fluid}$$

When a body is floating freely, it displaces a sufficient volume of fluid to just balance its own weight.

Figure 3–26 shows several devices which might be used in an undersea exploration project and which illustrate four different types of buoyancy problems. The buoy (a) and the ship (e) must be designed to float stably. The instrument package (b) would tend to float if not restrained by the anchoring cable. The diving bell (c) must be supported by the crane on a ship while the submarine (d) has the ability to hover at any depth (this is called neutral buoyancy).

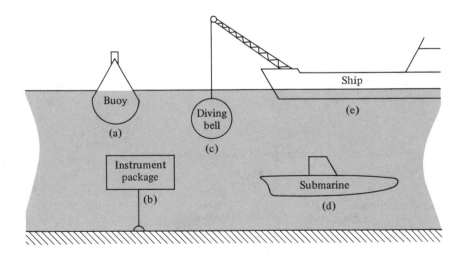

FIGURE 3–26

The analysis of problems dealing with buoyancy requires the application of the equation of static equilibrium in the vertical direction, $\Sigma F_v = 0$, assuming the object is at rest in the fluid. The procedure outlined below is recommended for all problems whether they involve floating or submerged bodies.

1. Decide what the objective of the problem solution is. Is it to find a force, weight, volume, or specific weight?

2. Draw a free body diagram of the object in the fluid. Show all forces which act on the free body in the vertical direction including the weight of the body, the buoyant force, and all external forces. If the direction of some force is not known, assume the most probable direction and show it on the free body.

3. Write the equation of static equilibrium in the vertical direction, $\Sigma F_v = 0$, assuming the positive direction to be upward.

4. Solve for the desired force, weight, volume, or specific weight, remembering the following concepts.

   a. The buoyant force is calculated from $F_b = \gamma_f V_d$.
   b. The weight of a solid object is the product of its total volume and its specific weight, that is, $w = \gamma V$.
   c. An object with an average specific weight less than that of the fluid will tend to float.
   d. An object with an average specific weight greater than that of the fluid will tend to sink.
   e. If the sign of an unknown force turns out to be negative, it means that the actual direction of the force is opposite to the direction assumed when drawing the free body diagram.

Two example problems will now be presented using the programmed instruction format. The solutions for two more problems are included which you can use to check your ability to solve buoyancy problems.

Programmed Example Problems

Example Problem 3–19. A cube, 18 inches on a side is made of bronze having a specific weight of 553 lb/ft³. Determine the magnitude and direction of the force required to hold the cube in equilibrium completely submerged (a) in water, and (b) in mercury. The specific gravity of mercury is 13.60.

Part (a) is considered first. Imagine the cube of bronze submerged in water. Now do Step 1 of the procedure.

Assuming that the bronze cube will not stay in equilibrium by itself, some external force is required. The objective is to find the magnitude of this force and the direction in which it would act, that is, up or down. Now do Step 2 of the procedure before looking at the next panel.

The free body is simply the cube itself. There are three forces acting on the cube in the vertical direction as shown in Fig. 3–27: the weight of the cube w, acting downward through its center of gravity; the buoyant force F_b, acting upward through the centroid of the displaced volume; and the externally applied supporting force F_e.

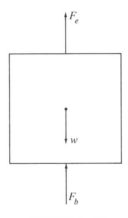

FIGURE 3–27

How do we know to draw the force F_e in the upward direction?

We really do not know for certain. However, experience should indicate that without an external force the solid bronze cube would tend to sink in water. Therefore, an upward force seems to be required to hold the cube in equilibrium. If our choice is wrong, the final result will indicate that to us.

Now assuming that the forces are as shown in Fig. 3–27, go on to Step 3.

The equation should look like this. (Assume that positive forces act upward.)

$$\Sigma F_v = 0$$
$$F_b + F_e - w = 0 \qquad \text{(3-12)}$$

As a part of Step 4, solve this equation algebraically for the desired term.

You should now have

$$F_e = w - F_b \qquad \text{(3-13)}$$

since the objective is to find the external force.

How do we calculate the weight of the cube w?

Item b under Step 4 of the procedure indicates that $w = \gamma_B V$, where γ_B is the specific weight of the bronze cube and V is its total volume. For the cube, since each side is 18 inches or 1.5 feet,

$$V = (1.5 \text{ ft})^3 = 3.38 \text{ ft}^3$$

and

$$w = \gamma_B V = \frac{553 \text{ lb}}{\text{ft}^3} \cdot 3.38 \text{ ft}^3 = 1870 \text{ lb}$$

There is another unknown on the right side of Eq. (3-13). How do we calculate F_b?

Check item (a) under Step 4 of the procedure if you have forgotten.

$$F_b = \gamma_f V_d$$

In this case γ_f is the specific weight of the water (62.4 lb/ft³) and the displaced volume V_d is equal to the total volume of the cube which we already know to be 3.38 ft³. Then,

$$F_b = \gamma_f V_d = \frac{62.4 \text{ lb}}{\text{ft}^3} \cdot 3.38 \text{ ft}^3 = 211 \text{ lb}$$

Now we can complete our solution for F_e.

The solution is

$$F_e = w - F_b = 1870 \text{ lb} - 211 \text{ lb} = 1659 \text{ lb}$$

Notice that the result is positive. This means that our assumed direction for F_e was correct. Then the solution to the problem is that an upward force of 1659 pounds is required to hold the block of bronze in equilibrium under water.

What about part (b) of the problem where the cube is submerged in mercury? Our objective is the same as before, to determine the magnitude and direction of the force required to hold the cube in equilibrium.

Now do Step 2 of the procedure.

Either of two free body diagrams are correct as shown in Fig. 3–28 depending on the assumed direction for the external force F_e. The solution for the two diagrams will be carried out simultaneously so you can check your work regardless of which diagram looks like yours and to demonstrate that either approach will yield the correct answer

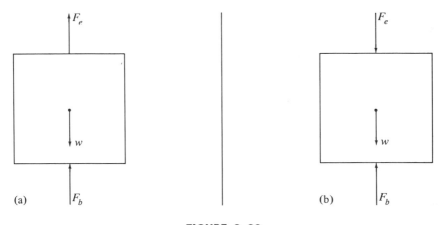

FIGURE 3–28

Now do Step 3 of the procedure.

These are the correct equations of equilibrium. Notice the differences and relate them to the figures.

$$F_b + F_e - w = 0 \qquad \Big| \qquad F_b - F_e - w = 0$$

Now solve for F_e.

You should now have

$$F_e = w - F_b \qquad \bigg| \qquad F_e = F_b - w$$

Since the magnitude of w and F_b are the same for each equation, they can now be calculated.

As in the preceding problem,

$$w = \gamma_B V = \frac{553 \text{ lb}}{\text{ft}^3} \cdot 3.38 \text{ ft}^3 = 1870 \text{ lb}$$

For the buoyant force F_b you should have

$$F_b = \gamma_m V = (sg)_m (62.4 \text{ lb/ft}^3)(V)$$

where the subscript, m, refers to mercury.

$$F_b = (13.60)\left(62.4 \frac{\text{lb}}{\text{ft}^3}\right)(3.38 \text{ ft}^3) = 2870 \text{ lb}$$

Now go on with the solution for F_e.

The correct answers are:

$F_e = w - F_b$	$F_e = F_b - w$
$F_e = 1870 \text{ lb} - 2870 \text{ lb}$	$F_e = 2870 \text{ lb} - 1870 \text{ lb}$
$F_e = -1000 \text{ lb}$	$F_e = +1000 \text{ lb}$

Notice that both solutions yield the same numerical value but they have opposite signs. The negative sign for the solution on the left means that the assumed direction for F_e in Fig. 3–28(a) was wrong. Therefore, both approaches give the same result. The required external force is a downward force of 1000 pounds.

How could you have reasoned from the start that a downward force would be required?

Items c and d of Step 4 of the procedure suggest that the specific weight of the cube and the fluid be compared. In this case:

For the bronze cube, $\gamma_B = 553 \text{ lb/ft}^3$

For the fluid (mercury), $\gamma_m = (13.60)(62.4 \text{ lb/ft}^3)$

$= 849 \text{ lb/ft}^3$

Since the specific weight of the cube is less than that of the mercury it would tend to float without an external force. Therefore, a downward force, as pictured in Fig. 3–28(b), would be required to hold it in equilibrium under the surface of the mercury.

This example problem is concluded.

Example Problem 3–20. A certain solid metal object has such an irregular shape that it is difficult to calculate its volume by geometry. The principle of buoyancy can be used to enable us to calculate its volume.

First the weight of the object is determined in the normal manner to be 60 pounds. Then using a setup similar to that in Fig. 3–29, its apparent weight while submerged in water is found to be 46.5 pounds. Using this data and the procedure for analyzing buoyancy problems we can find the volume of the object.

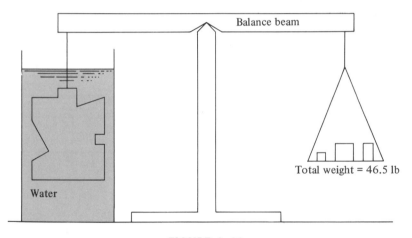

Balance beam

Total weight = 46.5 lb

Water

FIGURE 3–29

Now draw the free body diagram of the object while it is suspended in the water.

The free body diagram of the object while it is suspended in the water should look like Fig. 3–30. In this figure, what are the two forces F_e and w?

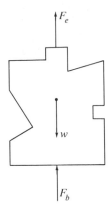

FIGURE 3–30

From the problem statement we should know that $w = 60$ lb, the weight of the object in air, and $F_e = 46.5$ lb, the supporting force exerted by the balance shown in Fig. 3–29.

Now do Step 3 of the procedure.

Using $\Sigma F_v = 0$, we get

$$F_b + F_e - w = 0$$

Our objective is to find the total volume of the object V. How can we get V from this equation?

This equation is used:

$$F_b = \gamma_f V$$

where γ_f is the specific weight of the water, 62.4 lb/ft³.

Substitute this into the preceding equation and solve for V.

You should now have

$$F_b + F_e - w = 0$$
$$\gamma_f V + F_e - w = 0$$
$$\gamma_f V = w - F_e$$
$$V = \frac{w - F_e}{\gamma_f}$$

Now we can put in the known values and calculate V.

The result is $V = 0.216$ ft³. This is how it is done.

$$V = \frac{w - F_e}{\gamma_f} = (60 - 46.5)\text{lb}\left(\frac{\text{ft}^3}{62.4 \text{ lb}}\right) = \frac{13.5 \text{ ft}^3}{62.4} = 0.216 \text{ ft}^3$$

Now that the volume of the object is known, the specific weight of the material can be found.

$$\gamma = \frac{w}{V} = \frac{60 \text{ lb}}{0.216 \text{ ft}^3} = 278 \text{ lb/ft}^3$$

This is approximately the specific weight of a titanium alloy.

The next two problems are worked out in detail and should serve to check your ability to solve buoyancy problems. After reading the problem statement, you should complete the solution yourself before reading the panel on which a correct solution is given. Be sure to read the problem carefully and to use the proper units in your calculations. Although there is more than one way to solve some problems, it is possible to get the correct answer by the wrong method. If your method is different from that given, be sure yours is based on sound principles before assuming it is correct.

Example Problem 3–21. A cube, 3 inches on a side, is made of a rigid foam material and floats in water with 2.5 inches below the surface. Calculate the magnitude and direction of the force required to hold it completely submerged in glycerine which has a specific gravity of 1.26.

Complete the solution before looking at the next panel.

Solution. Calculate the force required to hold the cube submerged in glycerine.

Free body diagrams, Fig. 3–31:

    a. Cube floating on water
    b. Cube submerged in glycerine

From Fig. 3–31(a),

$$\Sigma F_v = 0$$
$$F_b - w = 0$$
$$w = F_b = \gamma_f V_d$$
$$V_d = (3 \text{ in.})(3 \text{ in.})(2.5 \text{ in.}) = 22.5 \text{ in.}^3 \text{ (submerged}$$
$$\text{volume of cube)}$$
$$w = \frac{62.4 \text{ lb}}{\text{ft}^3} \cdot 22.5 \text{ in.}^3 \cdot \frac{\text{ft}^3}{1728 \text{ in.}^3} = 0.81 \text{ lb}$$

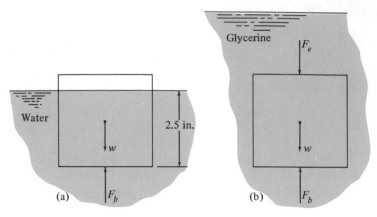

FIGURE 3-31

From Fig. 3–31(b),

$$\Sigma F_v = 0$$
$$F_b - F_e - w = 0$$
$$F_e = F_b - w = \gamma_f V_d - 0.81 \text{ lb}$$
$$V_d = (3 \text{ in.})(3 \text{ in.})(3 \text{ in.}) = 27 \text{ in.}^3 \text{ (total volume of cube)}$$
$$\gamma_f = (1.26)(62.4 \text{ lb/ft}^3) = 78.6 \text{ lb/ft}^3$$
$$F_e = \gamma_f V_d - 0.81 \text{ lb}$$
$$= \frac{78.6 \text{ lb}}{\text{ft}^3} \cdot 27 \text{ in.}^3 \cdot \frac{\text{ft}^3}{1728 \text{ in.}^3} - 0.81 \text{ lb}$$
$$F_e = 1.23 \text{ lb} - 0.81 \text{ lb} = 0.42 \text{ lb}$$

A downward force of 0.42 pounds is required to hold the cube submerged in glycerine.

Example Problem 3–22. A brass cube, 6 inches on a side, weighs 67 pounds. It is desired to hold this cube in equilibrium under water by attaching a light foam buoy to it. If the foam weighs 4.5 lb/ft³ what is the minimum required volume of the buoy?

Complete the solution yourself before looking at the next panel.

Solution. Calculate the minimum volume of foam to hold the brass cube in equilibrium.

Notice that the foam and brass in Fig 3–32 are considered as parts of a single system and that there is a buoyant force on each. The subscript F refers to the foam and the subscript B refers to the brass. No external force is required.

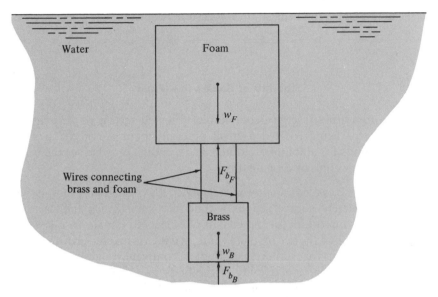

FIGURE 3-32

Equilibrium equation:

$$\Sigma F_v = 0$$
$$0 = F_{bB} + F_{bF} - w_B - w_F \tag{3-14}$$
$$w_B = 67 \text{ lb (given)}$$
$$F_{bB} = \gamma_f V_{dB} = \frac{62.4 \text{ lb}}{\text{ft}^3} \cdot (6 \text{ in.})^3 \cdot \frac{\text{ft}^3}{1728 \text{ in.}^3} = 7.8 \text{ lb}$$
$$w_F = \gamma_F V_F$$
$$F_{bF} = \gamma_f V_F$$

Substitute into Eq. (3–14),

$$F_{bB} + F_{bF} - w_B - w_F = 0$$
$$7.8 \text{ lb} + \gamma_f V_F - 67 \text{ lb} - \gamma_F V_F = 0$$

Solve for V_F, using $\gamma_f = 62.4 \text{ lb/ft}^3$, $\gamma_F = 4.5 \text{ lb/ft}^3$,

$$\gamma_f V_F - \gamma_F V_F = 67 \text{ lb} - 7.8 \text{ lb} = 59.2 \text{ lb}$$
$$V_F(\gamma_f - \gamma_F) = 59.2 \text{ lb}$$
$$V_F = \frac{59.2 \text{ lb}}{\gamma_f - \gamma_F} = \frac{59.2 \text{ lb ft}^3}{(62.4 - 4.5) \text{ lb}}$$
$$V_F = 1.02 \text{ ft}^3$$

This means that if 1.02 ft³ of foam is attached to the brass cube, the combination would be in equilibrium in water without any external force. It would be neutrally buoyant.

This completes the programmed example problems.

3–7   **Stability of Bodies in a Fluid**

A body in a fluid is considered stable if it will return to its original position after being rotated a small amount about a horizontal axis. The conditions for stability are different depending on whether the body is completely submerged or floating.

Stability of Completely Submerged Bodies. Two familiar examples of bodies completely submerged in a fluid are submarines and weather balloons. It is important for these kinds of objects to remain in a specific orientation despite the action of currents, winds, or maneuvering forces.

The condition for stability of bodies completely submerged in a fluid is that the center of gravity of the body must be below the center of buoyancy. The center of buoyancy of a body is at the centroid of the displaced volume of fluid and it is through this point that the buoyant force acts in a vertical direction. The weight of the body acts vertically downward through the center of gravity.

The undersea research vehicle shown in Fig. 3–33 has a stable configuration due to its shape and the location of equipment within the hull. Figure 3–34 shows the approximate cross-sectional shape of the vessel. The circular section is a hollow cylinder providing a manned cabin and housing for delicate instruments and life support systems. The rectangular section at the bottom contains heavy batteries and other durable equipment. With this distribution of weight and volume, the center of gravity (cg) and the center of buoyancy (cb) are located approximately as shown in Fig. 3–34(a).

Figure 3–34(b) shows the action of the buoyant force and the weight to produce a couple which tends to rotate the vessel back to its original position after being rotated slightly. Thus the body is stable. Contrast this with Fig. 3–34(c) which shows what would happen if the configuration were upside down from that in Fig. 3–34(a). When this body is rotated a small amount, the weight and the buoyant force produce a couple which tends to overturn it. Therefore, this orientation of the body is unstable.

If the center of gravity and center of buoyancy of a body coincide, as with a solid object, the weight and buoyant force act through the same point producing no couple. In this case the body would have

FIGURE 3–33

Grumman-BEN FRANKLIN, Undersea Research Vessel

Source of photo: Grumman Aerospace Corp.
Ocean Systems Department

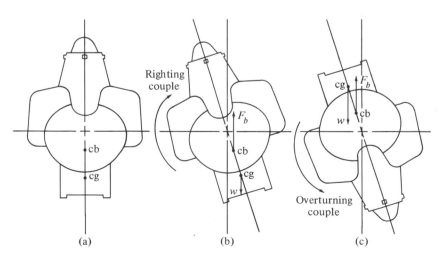

FIGURE 3–34

neutral stability and would remain in any orientation in which it is placed.

Stability of Floating Bodies. The conditions for stability of floating bodies is different from that for completely submerged bodies and the reason is illustrated in Fig. 3–35 which shows the approximate cross section of a ship's hull. In part (a) of the figure, the floating body is at its equilibrium orientation and the center of gravity (cg) is above the center of buoyancy (cb). A vertical line through these points will be called the vertical axis of the body. Figure 3–35 (b) shows that if the body is rotated slightly, the center of buoyancy shifts to a new position because the geometry of the displaced volume has changed. The buoyant force and the weight now produce a righting couple which tends to return the body to its original orientation. Thus the body is stable.

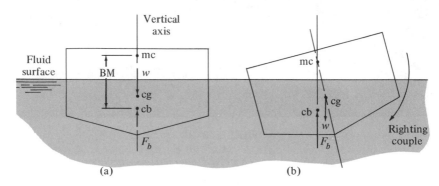

FIGURE 3–35

In order to state the condition for stability of a floating body, a new term, metacenter, must be defined. The metacenter (mc) is defined as the intersection of the vertical axis of a body when in its equilibrium position and a vertical line through the new position of the center of buoyancy when the body is rotated slightly. This is illustrated in Fig. 3–35(b). A floating body is stable if its center of gravity is below the metacenter.

It is possible to determine analytically if a floating body is stable by calculating the location of its metacenter. The distance from the center of buoyancy to the metacenter is called BM and is calculated from

$$\text{BM} = \frac{I}{V_d} \tag{3–15}$$

In this equation V_d is the displaced volume of fluid and I is the moment of inertia of a horizontal section of the body taken at the surface of the fluid. If the distance BM places the metacenter above the center of gravity, the body is stable.

Programmed Example Problems

Example Problem 3–23. Figure 3–36 (a) shows the cross-sectional shape of a flatboat hull and indicates the location of the center of gravity when fully loaded, weighing 32,000 pounds. Part (b) shows the top view of the boat. We would like to determine if the boat is stable.

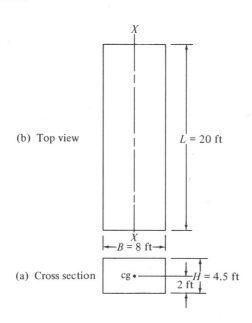

(b) Top view L = 20 ft

X

B = 8 ft

(a) Cross section   cg H = 4.5 ft
2 ft

FIGURE 3–36

Find out if the boat will float with 32,000 pounds in it.

This is done by finding how far the boat would sink into the water using the principles of buoyancy stated in Section 3–6. Complete that calculation before going to the next panel.

The draft of the boat is 3.20 feet, as shown in Fig. 3–37, found by the method described below. Equation of equilibrium: $\Sigma F_v = 0$

$$F_b - w = 0$$
$$w = F_b$$

Submerged volume: $V_d = B \cdot L \cdot X$
Buoyant force: $F_b = \gamma_f V_d = \gamma_f \cdot B \cdot L \cdot X$

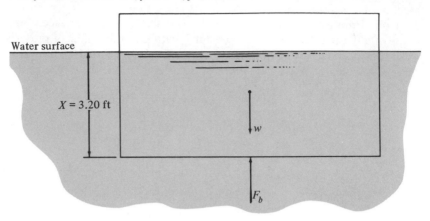

FIGURE 3–37

Then,

$$w = F_b = \gamma_f \cdot B \cdot L \cdot X$$

$$X = \frac{w}{B \cdot L \cdot \gamma_f} = \frac{32{,}000 \text{ lb}}{(8 \text{ ft})(20 \text{ ft})} \cdot \frac{\text{ft}^3}{62.4 \text{ lb}} = 3.20 \text{ ft}$$

It floats with 3.20 feet submerged. Where is the center of buoyancy?

It is at the center of the displaced volume of water. In this case, as shown in Fig. 3–38, it is on the vertical axis of the boat at a distance of 1.60 feet from the bottom. That is half of the draft, X.

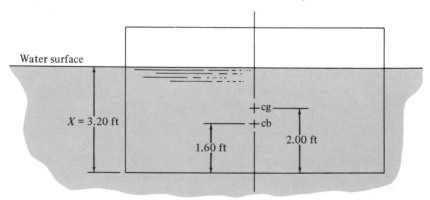

FIGURE 3–38

Since the center of gravity is above the center of buoyancy we must locate the metacenter to determine if the boat is stable. Using Eq. (3–15) calculate the distance BM and show it on the sketch.

The result is BM = 1.67 ft as shown in Fig. 3–39. Here is how it is done.

$$BM = I/V_d$$
$$V_d = L \cdot B \cdot X = (20 \text{ ft})(8 \text{ ft})(3.20 \text{ ft}) = 512 \text{ ft}^3$$

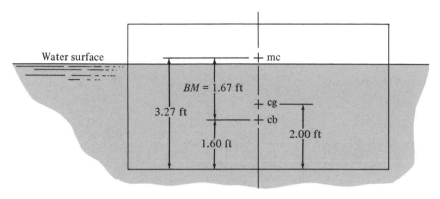

FIGURE 3–39

The moment of inertia I is determined about the axis X-X in Fig. 3–36 (b).

$$I = \frac{LB^3}{12} = \frac{(20 \text{ ft})(8 \text{ ft})^3}{12} = 853 \text{ ft}^4$$

Then,

$$BM = I/V_d = 853 \text{ ft}^4/512 \text{ ft}^3 = 1.67 \text{ ft}$$

Is the boat stable?

Yes, it is. Since the metacenter is above the center of gravity as shown in Fig. 3–39, the boat is stable.

Now read the next panel for another problem.

Example Problem 3–24. A solid cylinder is 3.0 feet in diameter, 6.0 feet high, and weighs 1550 pounds. If the cylinder is placed in oil (sg = 0.90), with its axis vertical, would it be stable?

The complete solution is shown in the next panel. Do this problem and then look at the solution.

Solution. Position of cylinder in oil: Fig. 3–40

$$V_d = \text{Submerged volume} = AX = \frac{\pi D^2}{4} \cdot X$$

Equilibrium equation: $\Sigma F_v = 0$

$$w = F_b = \gamma_o V_d = \gamma_o \frac{\pi D^2}{4} \cdot X$$

$$X = \frac{4w}{\pi D^2 \gamma_o} = \frac{(4)(1550 \text{ lb}) \text{ ft}^3}{(\pi)(3.0 \text{ ft})^2(0.90)(62.4 \text{ lb})} = 3.90 \text{ ft}$$

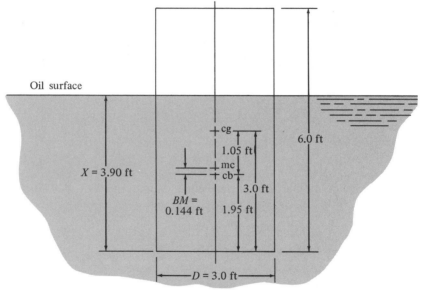

FIGURE 3–40

Center of buoyancy, cb, is at a distance $X/2$ from bottom of cylinder.

$$X/2 = 3.90 \text{ ft}/2 = 1.95 \text{ ft}$$

Center of gravity, cg, is at $H/2 = 3$ ft from bottom of cylinder assuming the material of the cylinder is of uniform specific weight. Position of metacenter, mc, using Eq. (3–15),

$$\text{BM} = I/V_d$$

$$I = \frac{\pi D^4}{64} = \frac{\pi(3 \text{ ft})^4}{64} = 3.98 \text{ ft}^4$$

$$V_d = AX = \frac{\pi D^2}{4} \cdot X = \frac{\pi(3 \text{ ft})^2}{4} \cdot 3.90 \text{ ft} = 27.6 \text{ ft}^3$$

$$\text{BM} = I/V_d = 3.98 \text{ ft}^4/27.6 \text{ ft}^3 = 0.144 \text{ ft}$$

Since this places the metacenter below the center of gravity, as shown in Fig. 3–40, the cylinder is not stable:

This completes the programmed instruction.

The conditions for stability of bodies in a fluid can be summarized as follows.

Completely submerged bodies are stable if the center of gravity is below the center of buoyancy.

Floating bodies are stable if the center of gravity is below the metacenter.

PRACTICE PROBLEMS

3–1 Express 84.5 psia as a gage pressure if the atmospheric pressure is 14.9 psia.

3–2 Express 22.8 psia as a gage pressure if the atmospheric pressure is 14.7 psia.

3–3 Express 4.3 psia as a gage pressure if the atmospheric pressure is 14.6 psia.

3–4 Express 10.8 psia as a gage pressure if the atmospheric pressure is 14.0 psia.

3–5 Express 14.7 psia as a gage pressure if the atmospheric pressure is 15.1 psia.

3–6 Express 41.2 psig as an absolute pressure if the atmospheric pressure is 14.5 psia.

3–7 Express 18.5 psig as an absolute pressure if the atmospheric pressure is 14.2 psia.

3–8 Express 0.6 psig as an absolute pressure if the atmospheric pressure is 14.7 psia.

3–9 Express −4.3 psig as an absolute pressure if the atmospheric pressure is 14.7 psia.

3–10 Express −12.5 psig as an absolute pressure if the atmospheric pressure is 14.4 psia.

3–11 At two points in a pipe line the pressures are 53.2 psig and 45.0 psig. Calculate the difference in pressure between the two points.

3–12   At a certain point A the pressure p_A is 32.4 psig. At a point B the pressure p_B is 28.6 psig. Calculate $p_A - p_B$.

3–13   At a point A the pressure p_A is 48.3 psig. At a point B the pressure p_B is 55.6 psig. Calculate $p_A - p_B$.

3–14   At a point 1 the pressure p_1 is 4.6 psi below atmospheric pressure. At a point 2 the pressure p_2 is 75.0 psig. Calculate $p_2 - p_1$.

3–15   At a point C the pressure p_C is 6.5 psia. At a point D the pressure p_D is 153.0 psig. Calculate $p_D - p_C$.

3–16   A storage tank for sulfuric acid is 1.5 feet in diameter and 4.0 feet high. If the acid has a specific gravity of 1.80, calculate the pressure at the bottom of the tank. The tank is open to the atmosphere at the top.

3–17   A storage drum for crude oil (sg = 0.89) is 32 feet deep and open at the top. Calculate the pressure at the bottom.

3–18   Figure 3–41 shows a closed container holding water and oil. Air at 5.0 psi below atmospheric pressure is above the oil. Calculate the pressure at the bottom of the container in psig.

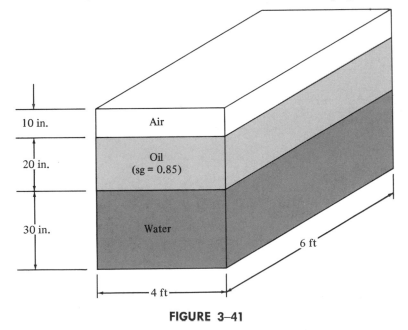

FIGURE 3–41

3–19 Determine the pressure at the bottom of the tank in Fig. 3–42.

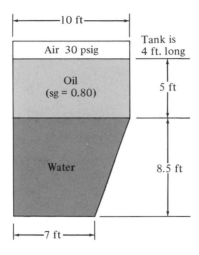

FIGURE 3–42

3–20 Determine the depth, h, of oil in the tank shown in Fig. 3–43.

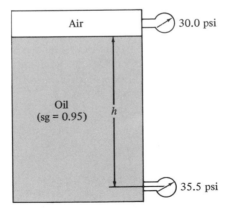

FIGURE 3–43

3–21 Figure 3–44 represents an oil storage drum which is open to the atmosphere at the top. Some water was accidentally pumped into the tank and settled to the bottom as shown in the figure. Calculate the depth of the water h_2, if the pressure gage at the bottom reads 23.0 psig.

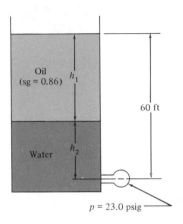

FIGURE 3–44

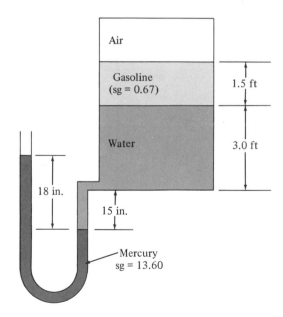

FIGURE 3–45

3–22 The greatest known depth in the ocean is approximately 36,000 feet. Assuming that the specific weight of the water is constant at 64.0 lb/ft³, calculate the pressure at this depth.

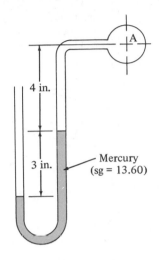

FIGURE 3–46

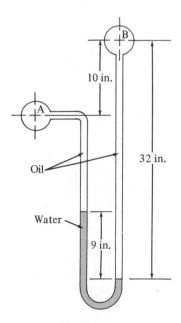

FIGURE 3–47

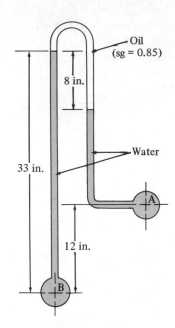

FIGURE 3–48

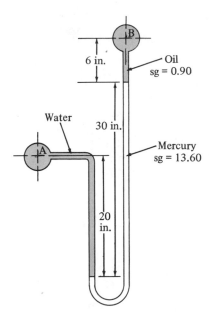

FIGURE 3–49

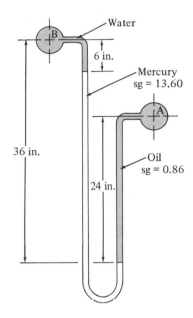

FIGURE 3–50

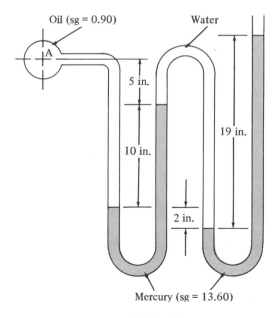

FIGURE 3–51

3–23   Figure 3–45 shows a closed tank containing gasoline floating on water. Calculate the air pressure above the gasoline.

3–24   Water is flowing in the pipe shown in Fig. 3–46. Calculate the pressure at point A in psig.

3–25   For the differential manometer shown in Fig. 3–47, calculate the pressure difference between points A and B. The specific gravity of the oil is 0.85.

3–26   For the manometer shown in Fig. 3–48 calculate $(p_A - p_B)$.

3–27   For the manometer shown in Fig. 3–49 calculate $(p_A - p_B)$.

3–28   For the manometer shown in Fig. 3–50 calculate $(p_A - p_B)$.

3–29   For the compound manometer shown in Fig. 3–51 calculate the pressure at point A.

3–30   For the compound differential manometer in Fig. 3–52 calculate $(p_B - p_A)$.

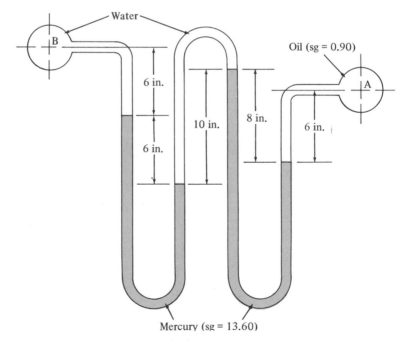

FIGURE 3–52

3–31 Figure 3–53 shows a manometer being used to indicate the difference in pressure between two points in a pipe. Calculate $(p_A - p_B)$.

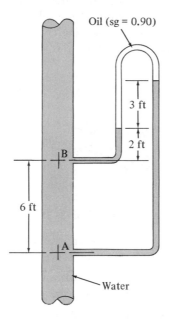

Oil (sg = 0.90)

3 ft

2 ft

B

6 ft

A

Water

FIGURE 3–53

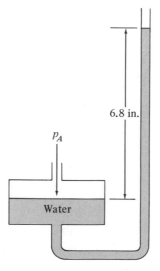

6.8 in.

p_A

Water

FIGURE 3–54

3–32   For the well-type manometer in Fig. 3–54 calculate p_A.

3–33   Figure 3–55 shows an inclined well-type manometer in which the distance L indicates the movement of the gage fluid level as the pressure p_A is applied above the well. The gage fluid has a specific gravity of 0.87 and $L = 4.6$ inches. Neglecting the drop in fluid level in the well, calculate p_A.

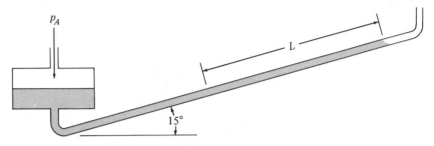

FIGURE 3–55

3–34   The barometric pressure is reported to be 28.6 inches of mercury. Calculate the atmospheric pressure in psia.

3–35   A barometer indicates the atmospheric pressure to be 30.65 inches of mercury. Calculate the atmospheric pressure in psia.

3–36   What would be the reading of a barometer in inches of mercury corresponding to an atmospheric pressure of 14.2 psia?

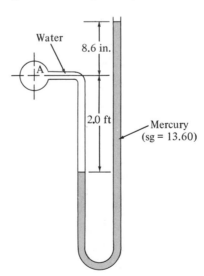

FIGURE 3–56

3-37 A barometer calibrated in the metric system of units reads 745 millimeters of mercury. Calculate the barometric reading in inches of mercury and the atmospheric pressure is psia.

3-38 (a) Determine the pressure at point *A* in Fig. 3-56 in psig. (b) If the barometric pressure is 29.0 inches of mercury, express the pressure at point *A* in psia.

3-39 Figure 3-57 shows a vacuum tank which has a flat circular observation window, 12 inches in diameter in one end. If the pressure in the tank is 0.12 psia when the barometer reads 30.5 inches of mercury, calculate the total force on the window.

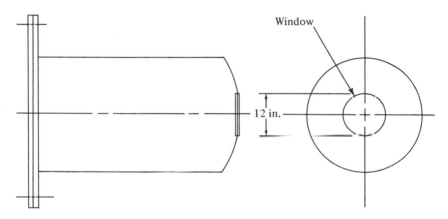

FIGURE 3-57

3-40 The flat end of the tank shown in Fig. 3-57 is secured with a bolted flange. If the inside diameter of the tank is 30 inches and the internal pressure is raised to +14.4 psig, calculate the total force which must be resisted by the bolts in the flange.

3-41 The egress hatch of a manned spacecraft is designed so that the internal pressure in the cabin applies a force to help maintain the seal. If the internal pressure is 5.0 psia and the external pressure is a perfect vacuum, calculate the force on a square hatch, 32 inches on a side.

3-42 Calculate the total force on the bottom of the closed tank shown in Fig. 3-41 if the air pressure is 7.5 psig.

3-43 If the length of the tank in Fig. 3-42 is 4 feet, calculate the total force on the bottom of the tank.

3-44   The wall shown in Fig. 3–58 is 20 feet long. (a) Calculate the total force on the wall due to water pressure and locate the center of pressure. (b) Calculate the moment due to this force at the base of the wall.

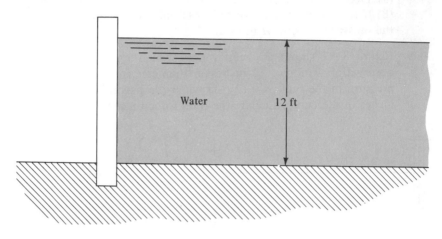

FIGURE 3–58

3-45   Calculate the total force on the wall in Fig. 3–59 due to the oil pressure if it is 12 feet long. Also determine the location of the center of pressure and show the resultant force on the wall.

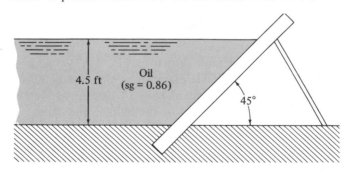

FIGURE 3–59

3-46   Figure 3–60 shows a gasoline tank filled into the filler pipe. The gasoline has a specific gravity of 0.67. Calculate the total force on each flat end of the tank and determine the location of the center of pressure.

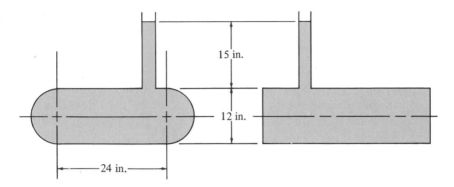

FIGURE 3–60

3–47   If the tank in Fig. 3–60 is filled just to the bottom of the filler pipe with gasoline (sg = 0.67) calculate the magnitude and location of the resultant force on the end.

3–48   If the tank in Fig. 3–60 is only half full of gasoline (sg = 0.67), calculate the magnitude and location of the resultant force on the end.

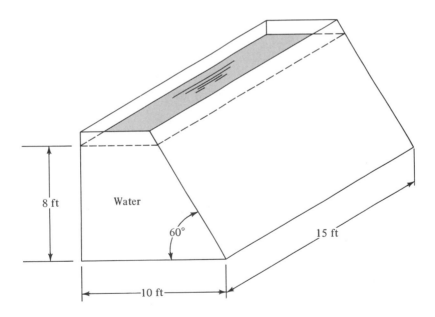

FIGURE 3–61

3–49   Figure 3–61 shows a water tank which is 15 feet long. Calculate the total force on (a) the bottom, (b) the vertical wall, (c) the inclined wall, and (d) each end wall.

3–50   Figure 3–62 shows the cross section of a reservoir for a hydraulic system. The length of the reservoir is 3.5 feet. Determine the magnitude and location of the total force on the side *AB* if the oil in the reservoir has a specific gravity of 0.93. Show the location and direction of the total force on a sketch.

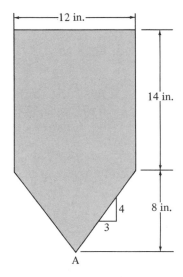

FIGURE 3–62

3–51   Figure 3–63 shows an oil storage tank. A circular observation port, 18 inches in diameter, is centered in the inclined portion of the tank. Calculate the magnitude and location of the resultant force exerted by the oil on the port.

3–52   A circular gate, 8 feet in diameter, is located in the inclined wall of the tank shown in Fig. 3–64. The tank contains a solution with a specific gravity of 1.10. Calculate the magnitude and location of the resultant force on the gate.

3–53   Calculate the magnitude and location of the total force on the vertical side of the tank in Fig. 3–64 if it is 10 feet long.

3–54   Calculate the magnitude and location of the total force on each end of the tank shown in Fig. 3–64 if they are vertical.

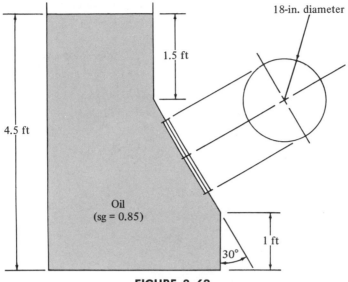

FIGURE 3–63

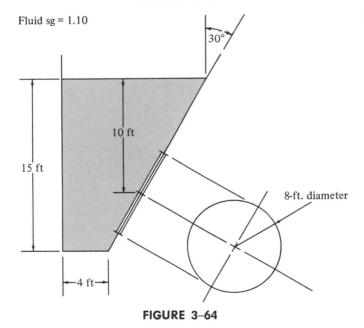

FIGURE 3–64

3–55 A reservoir for hydraulic oil has a circular inspection glass, 6 inches in diameter, mounted in an inclined panel as shown in Fig. 3–65. Calculate the magnitude and location of the resultant force on the glass.

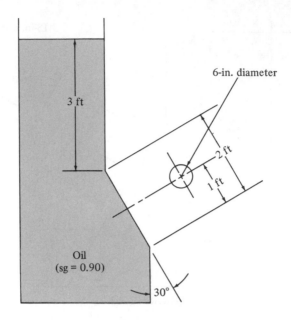

FIGURE 3–65

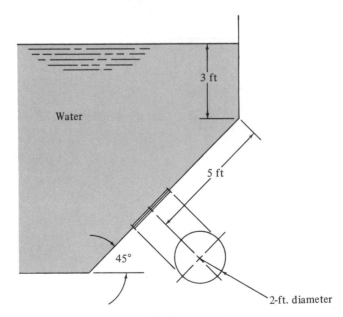

FIGURE 3–66

3–56 A circular observation window, 2.0 feet in diameter, is located as shown in Fig. 3–66 in the wall of a swimming pool. Calculate the magnitude and location of the resultant force on the window.

3–57 A door is located in the angled portion of the water tank shown in Fig. 3–67. Calculate the magnitude and location of the force on the door.

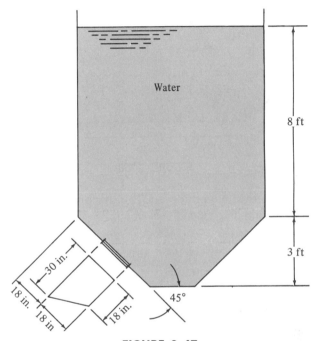

FIGURE 3–67

3–58 A window is located as shown in Fig. 3–68 in the inclined portion of the tank. Calculate the magnitude and location of the total force on the window.

3–59 A rectangular clean out door is located in the vertical back wall of a tank as shown in Fig. 3–68. (a) Calculate the magnitude and location of the resultant force on the door. (b) If the door is hinged at its top and has a latch at the bottom, calculate the force on the latch.

3–60 If the tank in Fig. 3–68 is 4 feet long, calculate the magnitude and location of the resultant force on the vertical end wall and the vertical rectangular back wall.

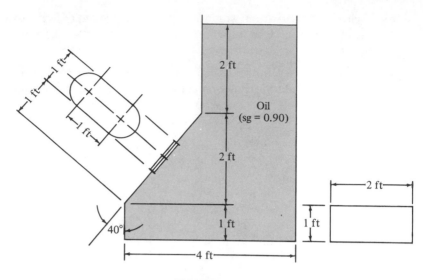

FIGURE 3–68

3–61   Determine the magnitude and direction of the resultant force
on the rectangular gate shown in Fig. 3–69.

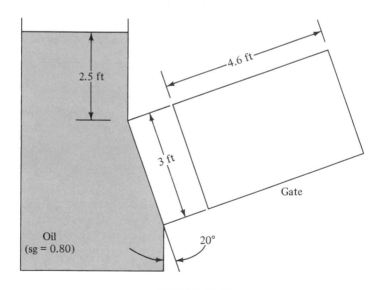

FIGURE 3–69

3–62   Calculate the resultant force on the inclined wall of the tank
shown in Fig. 3–42.

3–63 A steel cube, 4 inches on a side, weighs 18 pounds. It is desired to hold the cube in equilibrium under water by attaching a light foam buoy to it. If the foam weighs 3 lb/ft³ what is the minimum required volume of the buoy?

3–64 A concrete block having a specific weight of 150 lb/ft³ is suspended by a rope in a solution having a specific gravity of 1.15. What is the volume of the concrete block if the tension in the rope is 600 pounds?

3–65 Figure 3–70 shows a pump partially submerged in oil (sg = 0.90) and supported by springs. If the total weight of the pump is 14.6 pounds and the submerged volume is 40 in.³, calculate the supporting force exerted by the springs.

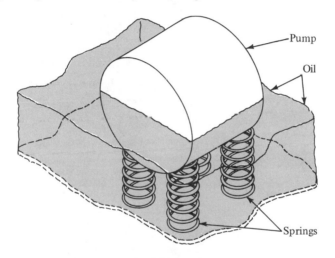

FIGURE 3–70

3–66 Figure 3–71 shows a cube floating in a fluid. Derive an expression relating the submerged depth X, the specific weight of the cube, and the specific weight of the fluid.

3–67 The instrument package shown in Fig. 3–72 weighs 58 pounds. Calculate the tension in the cable if the package is completely submerged in sea water having a specific weight of 64.0 lb/ft³.

3–68 A cylindrical drum is 2 feet in diameter, 3 feet long, and weighs 30 pounds when empty. Aluminum weights are to be placed inside the drum in order to make it neutrally buoyant in fresh water. What volume of aluminum will be required if it weighs 0.100 lb/in.³?

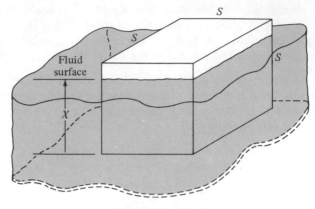

FIGURE 3–71

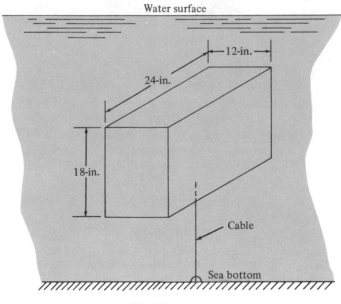

FIGURE 3–72

3–69   If the aluminum weights described in problem 3–68 are placed outside the drum, what volume will be required?

3–70   A 3.0 feet diameter hollow sphere weighing 45 pounds is attached to a solid concrete block weighing 920 pounds. If the concrete has a specific weight of 150 lb/ft³, will the two objects together float or sink in water?

3–71 A certain standard steel pipe has an outside diameter of 6.625 inches and a 1-foot length of the pipe weighs 19.0 pounds. Would the pipe float or sink in glycerine (sg = 1.26) if its ends are sealed?

3–72 A cylindrical float is 10 inches in diameter and 12 inches long. (a) What should be the specific weight of the float material if it is to have nine-tenths of its volume below the surface of a fluid with a specific gravity of 1.10? (b) If the cylinder is placed in the fluid with its axis vertical, would it be stable?

3–73 A buoy is a solid cylinder, 1.0 feet in diameter and 4 feet long. It is made of a material with a specific weight of 50 lb/ft³. (a) If it floats upright, how much of its length is above the water? (b) Is it stable?

3–74 A float to be used as a level indicator is being designed to float in oil which has a specific gravity of 0.90. It is to be a cube, 4 inches on a side and have 3 inches submerged in the oil. (a) Calculate the required specific weight of the float material. (b) Is the float stable?

3–75 A block of wood having a specific weight of 32 lb/ft³ has the dimensions 6 by 6 by 12 inches. If it is placed in oil (sg = 0.90) with the 6 by 12 inches surface parallel to the surface, would it be stable?

3–76 A barge is 60 feet long, 20 feet wide, and 8 feet deep. When empty it weighs 210,000 pounds and its center of gravity is 1.5 feet above the bottom. Is it stable when floating in water?

3–77 If the barge in problem 3–76 is loaded with 240,000 pounds of loose coal having an average specific weight of 45 lb/ft³, how much of the barge would be below the water? Is it stable?

3–78 A piece of cork having a specific weight of 15 lb/ft³ is shaped as shown in Fig. 3–73. (a) To what depth will it sink in turpentine (sg = 0.87) if placed in the orientation shown? (b) Is it stable in this position?

3–79 Figure 3–71 shows a cube floating in a fluid. (a) Derive an expression for the depth of submergence X which would insure that the cube is stable in the position shown. (b) Using the expression derived in (a) determine the required distance X for a cube 3 inches on a side.

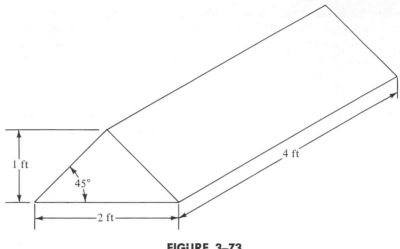

FIGURE 3–73

3–80   A boat has the cross section shown in Fig. 3–74(a). Its geometry at the water line is shown in the top view, Fig. 3–74(b). The hull is solid. Is the boat stable?

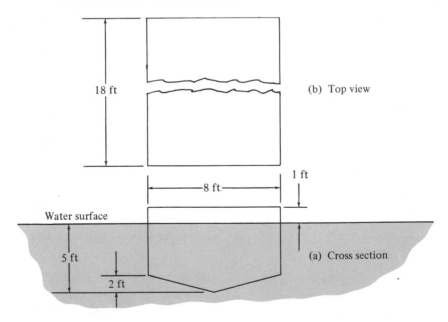

FIGURE 3–74

3–81 (a) If the cone shown in Fig. 3–75 is made of pine wood having a specific weight of 30 lb/ft³, would it be stable in the position shown floating in water? (b) Would it be stable if it is made of teak wood having a specific weight of 55 lb/ft³?

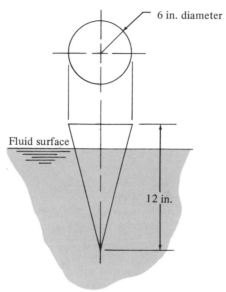

6 in. diameter

Fluid surface

12 in.

FIGURE 3–75

4

Flow of Fluids

Introduction

The material in the next seven chapters will be mainly concerned with the flow of fluids in closed circular pipes and tubes and with the devices used to control the flow. Applications are discussed involving fluid power systems, fluid distribution systems, pumps, turbines, valves, elbows, and other fittings. The techniques of analysis for many different situations are presented first. Then, in Chapters 9 and 10 these are combined in the solution of more complex pipe line problems.

When dealing with flow in pipes and tubes, we will assume that the fluid completely fills the available flow area unless otherwise stated. The analysis of open channel flow and flow in noncircular cross sections is presented in Chapter 11.

Most of the problems concerned with the flow of fluids in pipes and tubes involve the prediction of the conditions at one section in a system when the conditions at some other section are known. This is illustrated in Fig. 4–1 which shows a portion of a fluid distribution system with the flow going from section 1 at the bottom to Section 2 at the top. At any section in such a system we are usually concerned with the pressure of the fluid, the velocity of flow, and the elevation of the section. Remember, elevation is the term used to define the vertical distance from some reference level to a point of interest and is called z. When dealing with pipes and tubes the elevation is measured to the centerline of the section of interest.

At section 1 of the pipe in Fig. 4–1 we may know the pressure p_1, the velocity v_1, and the elevation z_1. Between sections 1 and 2 there may be pumps, valves, long lengths of pipe, elbows, changes in flow

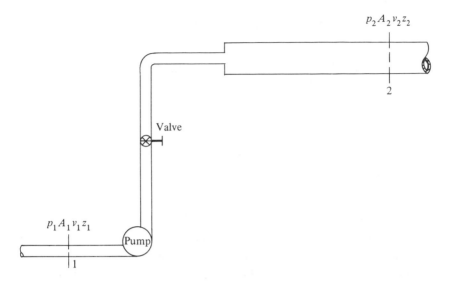

Figure 4–1

area, or many other things. Taking all these into account we would then determine the pressure p_2, the velocity v_2, and the elevation z_2.

4–2   **Fluid Flow Rate**

The quantity of fluid flowing in a system per unit time can be expressed by three different terms as defined below.

Q ft³/sec	The *volume flow rate* is the volume of fluid flowing past a section per unit time.
W lb/sec	The *weight flow rate* is the weight of fluid flowing past a section per unit time.
M slugs/sec	The *mass flow rate* is the mass of fluid flowing past a section per unit time.

The most fundamental of these three terms is the volume flow rate Q which is calculated from

$$Q = Av \qquad\qquad (4\text{–}1)$$

where A is the area of the section and v is the average velocity of flow. The units of Q can be derived as follows:

$$Q = Av = \text{ft}^2 \cdot \frac{\text{ft}}{\text{sec}} = \frac{\text{ft}^3}{\text{sec}}$$

The weight flow rate W is related to Q by

$$W = \gamma Q \qquad \qquad (4\text{--}2)$$

where γ is the specific weight of the fluid. The units of W are then

$$W = \gamma Q = \frac{\text{lb}}{\text{ft}^3} \cdot \frac{\text{ft}^3}{\text{sec}} = \frac{\text{lb}}{\text{sec}}$$

The mass flow rate M is related to Q by

$$M = \rho Q \qquad \qquad (4\text{--}3)$$

where ρ is the density of the fluid. The units of M are then

$$M = \rho Q = \frac{\text{slugs}}{\text{ft}^3} \cdot \frac{\text{ft}^3}{\text{sec}} = \frac{\text{slugs}}{\text{sec}}$$

The units given in Table 4–1 are in the English gravitational unit system which is used throughout this book. However, in many practical situations you will find other units in common use. For example, the volume flow rate is often expressed in gallons per minute (gpm), cubic centimeters per second, or liters per minute. The necessary conversion factors are included in the appendix.

Table 4–1 summarizes these three terms.

TABLE 4–1

Symbol	Name	Definition	Units
Q	Volume flow rate	$Q = Av$	ft³/sec
W	Weight flow rate	$W = \gamma Q$ $W = \gamma Av$	lb/sec
M	Mass flow rate	$M = \rho Q$ $M = \rho Av$	slugs/sec

4–3   **Continuity Equation**

The method of calculating the velocity of flow of a fluid in a closed pipe system depends on the principle of continuity. Consider the pipe in Fig. 4–2. A fluid is flowing from section 1 to section 2 at a constant rate. That is, the quantity of fluid flowing past any section in a given amount

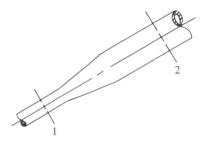

FIGURE 4–2

of time is constant. This is referred to as steady flow. Now if there is no fluid added, stored, or removed between section 1 and section 2, then the mass of fluid flowing past section 2 in a given amount of time must be the same as that flowing past section 1. This can be expressed in terms of the mass flow rate as

$$M_1 = M_2$$

or, since $M = \rho A v$,

$$\rho_1 A_1 v_1 = \rho_2 A_2 v_2 \qquad \qquad \textbf{(4–4)}$$

Equation (4–4) is a mathematical statement of the principle of continuity and is called the continuity equation. It is used to relate the fluid density, flow area, and velocity of flow at two sections of a system in which there is steady flow. It is valid for all fluids, either gases or liquids.

If the fluid in the pipe in Fig. 4–2 is a liquid which can be considered incompressible, then the terms ρ_1 and ρ_2 in Eq. (4–4) are equal. The equation then becomes

$$A_1 v_1 = A_2 v_2 \qquad \qquad \textbf{(4–5)}$$

or, since $Q = Av$,

$$Q_1 = Q_2$$

Equation (4–5) is the continuity equation as applied to liquids and states that for steady flow, the volume flow rate is the same at any section.

EXAMPLE PROBLEMS

Example Problem 4–1. In Fig. 4–2 the inside diameters of the pipe at sections 1 and 2 are 1.0 inch and 2.0 inches, respectively. Water at 160°F is flowing with an average velocity of 10 ft/sec at section 1. Calculate the following:

a) velocity at section 2
b) volume flow rate
c) weight flow rate
d) mass flow rate

Solution.

a) velocity at section 2.
   From Eq. (4–5),

$$A_1 v_1 = A_2 v_2$$

$$v_2 = v_1 \cdot \frac{A_1}{A_2}$$

$$A_1 = \frac{\pi D_1{}^2}{4} = \frac{\pi (1.0 \text{ in.})^2}{4} = 0.785 \text{ in.}^2$$

$$A_2 = \frac{\pi D_2{}^2}{4} = \frac{\pi (2.0 \text{ in.})^2}{4} = 3.14 \text{ in.}^2$$

Then,

$$v_2 = v_1 \cdot \frac{A_1}{A_2} = \frac{10 \text{ ft}}{\text{sec}} \cdot \frac{0.785 \text{ in.}^2}{3.14 \text{ in.}^2} = 2.5 \frac{\text{ft}}{\text{sec}}$$

Notice that for steady flow of a liquid, as the flow area increases, the velocity decreases. This is independent of pressure and elevation.

b) volume flow rate, Q.
   From Table 4–1, $Q = Av$. Because of the principle of continuity we could use the conditions either at section 1 or at section 2 to calculate Q. At section 1,

$$Q = A_1 v_1 = 0.785 \text{ in.}^2 \cdot \frac{10 \text{ ft}}{\text{sec}} \cdot \frac{1 \text{ ft}^2}{144 \text{ in.}^2} = 0.0545 \frac{\text{ft}^3}{\text{sec}}$$

c) weight flow rate, W.
   From Table 4–1, $W = \gamma Q$. At 160°F the specific weight of water is 61.0 lb/ft³. Then,

$$W = \gamma Q = \frac{61.0 \text{ lb}}{\text{ft}^3} \cdot \frac{0.0545 \text{ ft}^3}{\text{sec}} = 3.32 \frac{\text{lb}}{\text{sec}}$$

d) mass flow rate, M.
   From Table 4–1, $M = \rho Q$. At 160°F the density of water is 1.90 slugs/ft³. Then,

$$M = \rho Q = 1.90 \frac{\text{slugs}}{\text{ft}^3} \cdot \frac{0.0545 \text{ ft}^3}{\text{sec}} = 0.104 \frac{\text{slugs}}{\text{sec}}$$

Example Problem 4–2. At one section in an air distribution system, air at 14.7 psia and 100°F has an average velocity of 1200 ft/min and the duct is 12 inches square. At another section the duct is round with a diameter of 18 inches and the velocity is measured to be 900 ft/min. Calculate the density of the air in the round section and the weight flow rate of air in lb/hr.

Solution. According to the continuity equation for gases, Eq. (4–4),

$$\rho_1 A_1 v_1 = \rho_2 A_2 v_2$$

Then,

$$\rho_2 = \rho_1 \cdot \frac{A_1}{A_2} \cdot \frac{v_1}{v_2}$$

$$A_1 = (12 \text{ in.})(12 \text{ in.}) = 144 \text{ in.}^2$$

$$A_2 = \frac{\pi D_2^2}{4} = \frac{\pi (18 \text{ in.})^2}{4} = 254 \text{ in.}^2$$

At 14.7 psia and 100°F the density of air is 2.20×10^{-3} slugs/ft³. Then,

$$\rho_2 = 2.20 \times 10^{-3} \text{ slugs/ft}^3 \cdot \frac{144 \text{ in.}^2}{254 \text{ in.}^2} \cdot \frac{1200 \text{ ft/min}}{900 \text{ ft/min}}$$

$$\rho_2 = 1.66 \times 10^{-3} \text{ slugs/ft}^3$$

The weight flow rate can be found at section 1 from $W = \gamma_1 A_1 v_1$. At 14.7 psia and 100°F the specific weight of air is 7.09×10^{-2} lb/ft³. Then,

$$W = \gamma_1 A_1 v_1$$

$$W = 7.09 \times 10^{-2} \frac{\text{lb}}{\text{ft}^3} \cdot 144 \text{ in.}^2 \cdot \frac{1200 \text{ ft}}{\text{min}} \cdot \frac{1 \text{ ft}^2}{144 \text{ in.}^2} \cdot \frac{60 \text{ min}}{\text{hr}}$$

$$W = 5100 \text{ lb/hr}$$

4–4   **Conservation of Energy**

The analysis of a pipe line problem such as that illustrated in Fig. 4–1 involves an accounting for all of the energy within the system. In physics you learned that energy can neither be created nor destroyed but it can be transformed from one form to another. This is a statement of the law of conservation of energy.

There are three forms of energy which are always considered when analyzing a pipe flow problem. Consider an element of fluid as shown in Fig. 4–3 which may be inside a pipe in a flow system. It would be located at a certain elevation z, have a certain velocity v, and have a pressure p. The element of fluid would possess the following forms of energy.

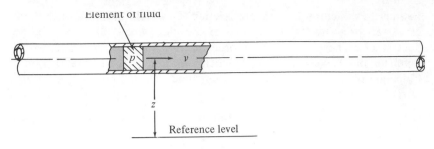

Element of fluid

Reference level

FIGURE 4–3

1. *Potential Energy.* Due to its elevation, the potential energy of the element relative to some reference level is

$$PE = wz \qquad (4\text{-}6)$$

where w is the weight of the element.

2. *Kinetic Energy.* Due to its velocity, the kinetic energy of the element is

$$KE = \tfrac{1}{2}\frac{wv^2}{g} \qquad (4\text{-}7)$$

3. *Flow Energy.* This is sometimes called pressure energy or flow work and represents the amount of work necessary to move the element of fluid across a certain section against the pressure p. Flow energy is abbreviated, FE, and is calculated from

$$FE = wp/\gamma \qquad (4\text{-}8)$$

Equation (4–8) can be derived as follows. Figure 4–4 shows the element of fluid in the pipe being moved across a section. The force on the element is pA where p is the pressure at the section and A is the area of the

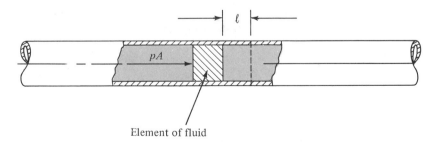

Element of fluid

FIGURE 4–4

section. In moving the element across the section, the force moves a distance ℓ equal to the length of the element. Therefore the work done is

$$\text{Work} = pA\ell = pV$$

where V is the volume of the element. The weight of the element w is

$$w = \gamma V$$

where γ is the specific weight of the fluid. Then

$$V = w/\gamma$$

and

$$\text{Work} = pV = pw/\gamma$$

which is called flow energy in Eq. (4–8).

The total amount of energy of these three forms possessed by the element of fluid would be the sum, called E.

$$E = \text{FE} + \text{PE} + \text{KE}$$
$$E = \frac{wp}{\gamma} + wz + \frac{wv^2}{2g}$$

Each of these terms has the units of energy which would be foot-pounds in the English gravitational unit system.

Now consider the element of fluid in Fig. 4–5 which moves from a section 1 to a section 2. The values for p, z, and v may be different at the two sections. At section 1 the total energy is

$$E_1 = \frac{wp_1}{\gamma} + wz_1 + \frac{wv_1^2}{2g}$$

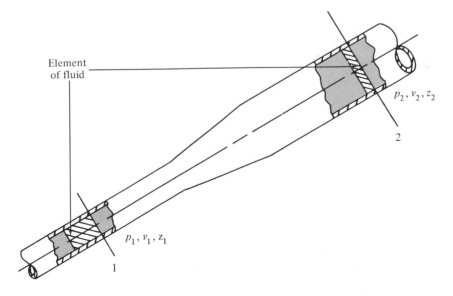

Element
of fluid

p_2, v_2, z_2

2

p_1, v_1, z_1

1

FIGURE 4–5

At section 2 the total energy is

$$E_2 = \frac{wp_2}{\gamma} + wz_2 + \frac{wv_2^2}{2g}$$

If no energy is added to the fluid or lost between sections 1 and 2, then the principle of conservation of energy requires that

$$E_1 = E_2$$
$$\frac{wp_1}{\gamma} + wz_1 + \frac{wv_1^2}{2g} = \frac{wp_2}{\gamma} + wz_2 + \frac{wv_2^2}{2g}$$

The weight of the element w is common to all terms and can be divided out. The equation then becomes

$$\frac{p_1}{\gamma} + z_1 + \frac{v_1^2}{2g} = \frac{p_2}{\gamma} + z_2 + \frac{v_2^2}{2g} \qquad (4\text{–}9)$$

This is referred to as Bernoulli's equation.

4–5   Interpretation of Bernoulli's Equation

A careful examination of Eq. (4–9) would indicate that each term has the unit of length which is feet in the English gravitational unit system. However, a more reasonable set of units is *foot-pounds per pound*. This is dimensionally the same as feet, but its physical interpretation is that each term represents a portion of the energy possessed by the fluid, per pound of fluid flowing in the system.

Because the terms in Bernoulli's equation have the unit of feet, they are often spoken of as "head" referring to a height above some reference level. The term p/γ is called *pressure head; z* is *elevation head;* and $v^2/2g$ is *velocity head*. The sum of these three is the *total head*.

4–6   Restrictions on Bernoulli's Equation

Although Bernoulli's equation is applicable to a large number of practical problems, there are several limitations which must be understood in order to apply it properly.

1. It is valid only for incompressible fluids since the specific weight of the fluid is assumed to be the same at the two sections of interest.

2. There can be no mechanical devices between the two sections of interest which would add or remove energy from the system, since the equation states that the total energy in the fluid is constant.
3. There can be no heat transferred into or out of the fluid.
4. There can be no energy lost due to friction.

In reality no system satisfies all of these restrictions. However, there are many systems for which only a negligible error will result by using Bernoulli's equation. Also, the use of this equation may allow a fast estimate of a result when that is all that is required.

4-7	**Applications of Bernoulli's Equation**

Several programmed example problems are presented below to illustrate the use of Bernoulli's equation. Although it is not possible to cover all types of problems with a certain solution method, the general approach to problems of fluid flow is described here.

1. Decide which items are known and what is to be found.
2. Decide which two sections in the system will be used when writing Bernoulli's equation. One section is chosen for which much data is known. The second is usually the section at which something is to be calculated.
3. Write Bernoulli's equation for the two selected sections in the system.
4. Solve the equation algebraically for the desired term.
5. Substitute known quantities and calculate the result.

Programmed Example Problems

Example Problem 4-3. In Fig. 4–6 water at 50°F is flowing from section 1 to section 2. At section 1, which is 1.0 inch in diameter, the pressure is 50.0 psig and the velocity of flow is 10 ft/sec. Section 2, which is 2.0 inches in diameter, is 6.5 feet above section 1. Assuming there are no energy losses in the system, calculate the pressure p_2.

List the items that are known from the problem statement before looking at the next panel.

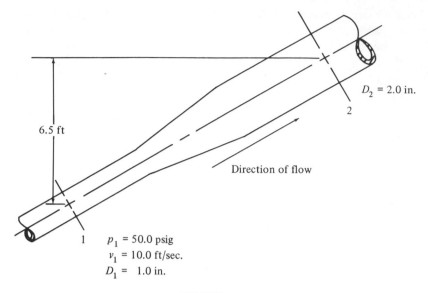

6.5 ft

D_2 = 2.0 in.

2

Direction of flow

1   p_1 = 50.0 psig
     v_1 = 10.0 ft/sec.
     D_1 = 1.0 in.

FIGURE 4–6

$D_1 = 1.0$ in. $v_1 = 10$ ft/sec $z_2 - z_1 = 6.5$ ft
$D_2 = 2.0$ in. $p_1 = 50$ psig

The pressure p_2 is to be found. In other words, we are asked to calculate the pressure at section 2 which is different from the pressure at section 1 because there is a change in elevation and flow area between the two sections.

We are going to use Bernoulli's equation to solve the problem. Which two sections should be used when writing the equation?

In this case sections 1 and 2 are the obvious choices. At section 1 we know p_1, v_1, and z_1. The unknown pressure p_2 is at section 2.

Now write Bernoulli's equation. [See Eq. (4–9).]

It should look like this:

$$\frac{p_1}{\gamma} + z_1 + \frac{v_1{}^2}{2g} = \frac{p_2}{\gamma} + z_2 + \frac{v_2{}^2}{2g}$$

The three terms on the left refer to section 1 while the three on the right refer to section 2.

Solve for p_2 in terms of the other variables.

The algebraic solution for p_2 could look like this.

$$\frac{p_1}{\gamma} + z_1 + \frac{v_1^2}{2g} = \frac{p_2}{\gamma} + z_2 + \frac{v_2^2}{2g}$$

$$\frac{p_2}{\gamma} = \frac{p_1}{\gamma} + z_1 + \frac{v_1^2}{2g} - z_2 - \frac{v_2^2}{2g}$$

$$p_2 = \gamma \left[\frac{p_1}{\gamma} + z_1 + \frac{v_1^2}{2g} - z_2 - \frac{v_2^2}{2g} \right]$$

This is correct. However, it is convenient to group the elevation heads and velocity heads together. Also, since $\gamma(p_1/\gamma) = p_1$, the final solution for p_2 should be

$$p_2 = p_1 + \gamma \left[z_1 - z_2 + \frac{v_1^2 - v_2^2}{2g} \right] \qquad \textbf{(4–10)}$$

Are the values of all the terms on the right side of this equation known?

Everything was given except γ, v_2, and g. Of course, $g = 32.2$ ft/sec². Since water at 50°F is flowing in the system, $\gamma = 62.4$ lb/ft³. How can v_2 be determined?

The continuity equation is used.

$$A_1 v_1 = A_2 v_2$$

$$v_2 = v_1 (A_1/A_2)$$

Calculate v_2 now.

You should have $v_2 = 2.5$ ft/sec found from

$A_1 = \pi D_1^2/4 = \pi(1.0 \text{ in.})^2/4 = 0.785$ in.²

$A_2 = \pi D_2^2/4 = \pi(2.0 \text{ in.})^2/4 = 3.14$ in.²

$v_2 = v_1 (A_1/A_2) = 10$ ft/sec $(0.785 \text{ in.}^2/3.14 \text{ in.}^2) = 2.5$ ft/sec

Now substitute the known values into Eq. (4–10).

$$p_2 = \frac{50.0 \text{ lb}}{\text{in.}^2} + \frac{62.4 \text{ lb}}{\text{ft}^3}\left[-6.5 \text{ ft} + \frac{(10 \text{ ft/sec})^2 - (2.5 \text{ ft/sec})^2}{2\,(32.2 \text{ ft/sec}^2)}\right]$$

Notice that $z_1 - z_2 = -6.5$ ft. Neither z_1 nor z_2 is known but it is known that z_2 is 6.5 feet greater than z_1. Therefore the difference $z_1 - z_2$ must be negative.

Now complete the calculation for p_2 in the units of lb/in.^{2}

The final answer is $p_2 = 47.8$ psig. This is 2.2 psi less than p_1. The details of the solution are:

$$p_2 = 50.0\,\frac{\text{lb}}{\text{in.}^2} + 62.4\,\frac{\text{lb}}{\text{ft}^3}\left[-6.5 \text{ ft} + \frac{(100 - 6.25)}{(2)\,(32.2)}\,\frac{\text{ft}^2}{\text{sec}^2}\,\frac{\text{sec}^2}{\text{ft}}\right]$$

$$= 50.0\,\frac{\text{lb}}{\text{in.}^2} + 62.4\,\frac{\text{lb}}{\text{ft}^3}\,[-6.5 \text{ ft} + 1.46 \text{ ft}]\,\frac{1 \text{ ft}^2}{144 \text{ in.}^2}$$

$$= 50.0\,\frac{\text{lb}}{\text{in.}^2} - 2.2\,\frac{\text{lb}}{\text{in.}^2} = 47.8\,\frac{\text{lb}}{\text{in.}^2}$$

$$p_2 = 47.8 \text{ psig}$$

Another example of the use of Bernoulli's equation follows.

Example Problem 4–4. Figure 4–7 shows a nozzle attached to a pipe which has an internal diameter of 3.0 inches. The stream of water exiting from the nozzle is 2.0 inches in diameter. If the pressure in the pipe just ahead of the nozzle is 150 psig, calculate the volume flow rate of water through the nozzle in gallons per minute.

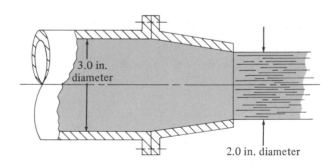

FIGURE 4–7

List the items which are known from the problem statement before looking at the next panel.

These items were given:

> Pressure ahead of nozzle = 150 psig.
> Pipe diameter = 3.0 in.
> Nozzle jet diameter = 2.0 in.
> The fluid is water. Therefore, $\gamma = 62.4$ lb/ft³.

What is the pressure in the free stream of fluid just outside the nozzle?

It is atmospheric pressure, 0 psig. Since there is no rigid pipe surrounding the stream, it must have the same pressure as the air around it.

Now, which two points should be used when writing Bernoulli's equation to solve this problem?

The only two points for which sufficient data is known are just ahead of the nozzle and in the free stream of water just outside the nozzle as shown in Fig. 4–8. At these points,

$$p_1 = 150 \text{ psig} \qquad p_2 - 0 \text{ psig}$$
$$D_1 = 3.0 \text{ in.} \qquad D_2 = 2.0 \text{ in.}$$

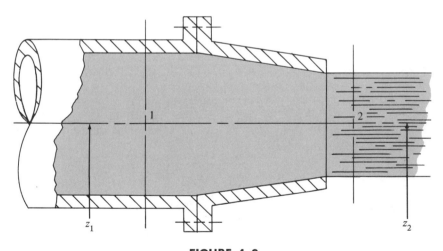

FIGURE 4–8

Also, since the nozzle is shown to be horizontal, the elevations z_1 and z_2 are equal.

Now write Bernoulli's equation for these two points.

You should have

$$\frac{p_1}{\gamma} + z_1 + \frac{v_1^2}{2g} = \frac{p_2}{\gamma} + z_2 + \frac{v_2^2}{2g}$$

This equation can be simplified quite a bit. Since $z_1 = z_2$, they cancel from the equation. Since $p_2 = 0$ psig it can be eliminated. We are left with

$$\frac{p_1}{\gamma} + \frac{v_1^2}{2g} = \frac{v_2^2}{2g} \qquad \text{(4–11)}$$

The objective of our solution is to determine the volume flow rate of water which is dependent on the velocity. However, at this point v_1 and v_2 are unknown. In order to calculate the velocities we need to use the continuity equation, which relates v_1 and v_2.

$$A_1 v_1 = A_2 v_2 \qquad \text{(4–12)}$$

This equation combined with Eq. (4–11) makes it possible to solve for one of the velocities, v_1 or v_2. First solve Eq. (4–12) for v_2 in terms of v_1.

$$v_2 = v_1 \, (A_1/A_2)$$

Evaluating A_1 and A_2 gives

$$A_1 = \pi D_1^2/4 = \pi(3.0 \text{ in.})^2/4 = 7.07 \text{ in.}^2$$
$$A_2 = \pi D_2^2/4 = \pi(2.0 \text{ in.})^2/4 = 3.14 \text{ in.}^2$$

Then,

$$v_2 = v_1 \, \frac{A_1}{A_2} = v_1 \cdot \frac{7.07 \text{ in.}^2}{3.14 \text{ in.}^2} = 2.25 \, v_1$$

This means that the term $2.25 \, v_1$ can be substituted into Eq. (4–11) in place of v_2. We then get

$$\frac{p_1}{\gamma} + \frac{v_1^2}{2g} = \frac{(2.25 \, v_1)^2}{2g} \qquad \text{(4–13)}$$

The only unknown in this equation is v_1. Solve for v_1 now.

The algebraic solution could look like this.

$$\frac{p_1}{\gamma} + \frac{v_1^2}{2g} = \frac{(2.25 \, v_1)^2}{2g} = \frac{5.06 \, v_1^2}{2g}$$
$$\frac{p_1}{\gamma} = \frac{5.06 \, v_1^2}{2g} - \frac{v_1^2}{2g} = \frac{4.06 \, v_1^2}{2g}$$

Then,

$$v_1 = \sqrt{\frac{2gp_1}{4.06\ \gamma}}$$

Now calculate v_1 in ft/sec.

The correct answer is $v_1 = 74.1$ ft/sec. Now calculate Q.

From $Q = A_1 v_1$ we get

$$Q = 7.07 \text{ in.}^2 \cdot \frac{74.1 \text{ ft}}{\text{sec}} \cdot \frac{\text{ft}^2}{144 \text{ in.}^2} = 3.64 \text{ ft}^3/\text{sec}$$

Converting to gallons per minute,

$$Q = 3.64 \text{ ft}^3/\text{sec} \cdot \frac{449 \text{ gal/min}}{1 \text{ ft}^3/\text{sec}} = 1630 \text{ gal/min}$$

This problem is now completed. The important feature here is that Bernoulli's equation and the continuity equation were both used to solve for one of the unknown velocities.

Example Problem 4–5. Figure 4–9 shows a siphon being used to draw water from a swimming pool. The pipe which makes up the siphon is a standard Schedule 40, $1\frac{1}{2}$ inch steel pipe which terminates with a 1.0 inch diameter nozzle. Assuming that there are no energy losses in the system, calculate the volume flow rate through the siphon and the pressure at points B, C, D, and E.

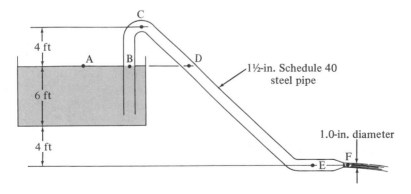

FIGURE 4–9

The first step in this problem solution is to calculate the volume flow rate, Q, using Bernoulli's equation. The two most convenient points to use for this calculation are A and F. What is known about point A?

Point A is the free surface of the water in the pool. Therefore $p_A = 0$ psig. Also, since the surface area of the pool is very large the velocity of the water at the surface is very nearly zero. Therefore, we will assume $v_A = 0$.

What do we know about point F?

Point F is in the free stream of water outside the nozzle. Since this stream is exposed to atmospheric pressure, the pressure $p_F = 0$ psig. We also know that point F is 10 feet below point A.

Now write Bernoulli's equation for points A and F.

You should have

$$\frac{p_A}{\gamma} + z_A + \frac{v_A^2}{2g} = \frac{p_F}{\gamma} + z_F + \frac{v_F^2}{2g}$$

Taking into account the information in the previous two panels, how can we simplify this equation?

Since $p_A = 0$ psig, $p_F = 0$ psig, and v_A is approximately zero, we can cancel them from the equation. What remains is this:

$$z_A = z_F + \frac{v_F^2}{2g}$$

The objective is to calculate the volume flow rate which depends on the velocity v_F. Solve for v_F now.

$$v_F = \sqrt{(z_A - z_F)2g}$$

What is $(z_A - z_F)$?

From Fig. 4–9 we see that $z_A - z_F = 10$ ft. Notice that the difference is positive since z_A is greater than z_F. We can now solve for the value of v_F in ft/sec.

The result is $v_F = 25.4$ ft/sec.

$$v_F = \sqrt{(10 \text{ ft})(2)(32.2 \text{ ft/sec}^2)} = \sqrt{644} \text{ ft/sec} = 25.4 \text{ ft/sec}$$

Now, how can Q be calculated?

Using the continuity equation, $Q = Av$, with the units of ft³/sec for Q, complete the calculation.

The result is $Q = 0.139$ ft³/sec.

$$Q = A_F v_F$$
$$v_F = 25.4 \text{ ft/sec}$$
$$A_F = \pi(1.0 \text{ in.})^2/4 = 0.785 \text{ in.}^2$$
$$Q = 0.785 \text{ in.}^2 \cdot \frac{25.4 \text{ ft}}{\text{sec}} \cdot \frac{\text{ft}^2}{144 \text{ in.}^2} = 0.139 \text{ ft}^3/\text{sec}$$

The first part of the problem is now complete. Now use Bernoulli's equation to determine p_B. What two points should be used?

Points A and B are the best. As shown in the previous panels, using point A allows the equation to be simplified greatly. And since we are looking for p_B, we must choose point B.

Write Bernoulli's equation for points A and B, simplify it as before, and solve for p_B.

Here is one possible solution procedure.

$$\frac{p_A}{\gamma} + z_A + \frac{v_A^2}{2g} = \frac{p_B}{\gamma} + z_B + \frac{v_B^2}{2g}$$

Since $p_A = 0$ psig and $v_A = 0$,

$$z_A = \frac{p_B}{\gamma} + z_B + \frac{v_B^2}{2g}$$
$$p_B = \gamma[(z_A - z_B) - v_B^2/2g] \qquad \text{(4-14)}$$

What is $z_A - z_B$?

It is zero. Since the two points are on the same level their elevations are the same. Can v_B be found?

---

v_B can be calculated using the continuity equation.

$$Q = A_B v_B$$
$$v_B = Q/A_B$$

The area of a standard Schedule 40, $1\frac{1}{2}$-inch steel pipe can be found in the appendix. Complete the calculation for v_B.

The result is $v_B = 9.8$ ft/sec.

$$v_B = Q/A_B$$
$$Q = 0.139 \text{ ft}^3/\text{sec}$$
$$A_B = 0.01414 \text{ ft}^2$$
$$v_B = \frac{0.139 \text{ ft}^3}{\text{sec}} \cdot \frac{1}{0.01414 \text{ ft}^2} = 9.8 \text{ ft/sec}$$

We now have all the data we need to calculate p_B from Eq. (4–14).

$$p_B = \gamma[(z_A - z_B) - v_B^2/2g]$$
$$\frac{v_B^2}{2g} = \frac{(9.8)^2 \text{ ft}^2}{\text{sec}^2} \cdot \frac{\text{sec}^2}{(2)(32.2) \text{ ft}} = 1.49 \text{ ft}$$
$$p_B = \frac{62.4 \text{ lb}}{\text{ft}^3}[0 - 1.49 \text{ ft}]$$
$$p_B = -92.9 \frac{\text{lb}}{\text{ft}^2} \cdot \frac{1 \text{ ft}^2}{144 \text{ in.}^2}$$
$$p_B = -0.645 \text{ lb/in.}^2$$

The negative sign indicates that p_B is 0.645 psi below atmospheric pressure. Notice that when dealing with fluids in motion, the concept that points on the same level have the same pressure does *not* apply as it does with fluids at rest.

The next three panels present the solutions for the pressures p_C, p_D, and p_E which can be found in a manner very similar to that used for p_B. Complete the solution for p_C yourself before looking at the next panel.

The answer is $p_C = -2.38$ psig.
Bernoulli's equation:

$$\frac{p_A}{\gamma} + z_A + \frac{v_A^2}{2g} = \frac{p_C}{\gamma} + z_C + \frac{v_C^2}{2g}$$

Since $p_A = 0$ and $v_A = 0$,

$$z_A = \frac{p_C}{\gamma} + z_C + \frac{v_C^2}{2g}$$

$$p_C = \gamma[(z_A - z_C) - v_C^2/2g]$$

$$z_A - z_C = -4 \text{ ft (negative since } z_C \text{ is greater than } z_A)$$

$$v_C = v_B = 9.8 \text{ ft/sec (since } A_C = A_B.)$$

$$\frac{v_C^2}{2g} = \frac{v_B^2}{2g} = 1.49 \text{ ft}$$

$$p_C = \frac{62.4 \text{ lb}}{\text{ft}^3}[-4 \text{ ft} - 1.49 \text{ ft}]$$

$$p_C = \frac{-342 \text{ lb}}{\text{ft}^2} \cdot \frac{1 \text{ ft}^2}{144 \text{ in.}^2} = \frac{-2.38 \text{ lb}}{\text{in.}^2}$$

$$p_C = -2.38 \text{ psig}$$

Complete the calculation for p_D before looking at the next panel.

The answer is $p_D = -0.645$ psig. This is the same as p_B since the elevation and the velocity at points B and D are equal. Solution by Bernoulli's equation would prove this.

Now find p_E.

At point E, $p_E = 3.68$ psig.
Bernoulli's equation:

$$\frac{p_A}{\gamma} + z_A + \frac{v_A^2}{2g} = \frac{p_E}{\gamma} + z_E + \frac{v_E^2}{2g}$$

Since $p_A = 0$ and $v_A = 0$,

$$z_A = \frac{p_E}{\gamma} + z_E + \frac{v_E^2}{2g}$$

$$p_E = \gamma[(z_A - z_E) - v_E^2/2g]$$

$$z_A - z_E = +10 \text{ ft}$$

$$v_E = v_B = 9.8 \text{ ft/sec}$$

$$\frac{v_E^2}{2g} = \frac{v_B^2}{2g} = 1.49 \text{ ft}$$

$$p_E = \frac{62.4 \text{ lb}}{\text{ft}^3}[10 \text{ ft} - 1.49 \text{ ft}]$$

$$p_E = \frac{531 \text{ lb}}{\text{ft}^2} \cdot \frac{1 \text{ ft}^2}{144 \text{ in.}^2} = \frac{3.68 \text{ lb}}{\text{in.}^2}$$

$$p_E = 3.68 \text{ psig}$$

4–8   Commercially Available Pipe and Tubing

The actual outside and inside diameters of standard commercially available pipe and tubing may be quite different from the nominal size given. Several widely used types of standard pipe and tubing will be described in this section. Data are given in the appendix for outside diameter, inside diameter, wall thickness, and flow area for these types.

Steel Pipe. General purpose pipe lines are often constructed of steel pipe. Standard pipe sizes are designated by the nominal size and schedule number. Schedule numbers are related to the permissible operating pressure of the pipe and to the allowable stress of the steel in the pipe. The range of schedule numbers is from 10 to 160, with the higher numbers indicating a heavier wall thickness. Since all schedules of pipe of a given nominal size have the same outside diameter, the higher schedules have a smaller inside diameter. The most complete series of steel pipe available are Schedules 40, 80, and 160. Data for these are given in the appendix.

Steel Tubing. Standard steel tubing is used in hydraulic systems, condensers, heat exchangers, engine fuel systems, and industrial fluid processing systems. Sizes are designated by outside diameter and wall thickness. Standard sizes from $\frac{1}{8}$ inch to 2 inches for several wall thickness gauges are tabulated in the appendix.

Copper Tubing. Household plumbing, refrigerant lines, and compressed air lines often use copper tubing manufactured as Type K or Type L. Type K has the greater wall thickness of the two and is recommended for underground service. Type L is suitable for general purpose interior plumbing. The nominal size of copper tubing is $\frac{1}{8}$ inch less than the actual outside diameter of the tube. Data for wall thickness, inside diameter, and flow area are given in the appendix.

Other Types of Pipe and Tubing. Water, gas, and sewage lines are often made of cast iron pipe. Brass pipe is used for corrosive fluids as is stainless steel. Other materials used are aluminum, lead, tin, vitrified clay, concrete, and many types of plastics such as polyethylene, nylon, and polyvinyl chloride.

PRACTICE PROBLEMS

In the following problems you may be required to refer to the appendix for fluid properties, dimensions of pipe and tubing, or conversion factors. Assume that there are no energy losses in all problems. Unless otherwise stated, the pipe sizes given are actual inside diameters.

4–1 Convert a volume flow rate of 3.0 gallons per minute to ft³/sec.

4–2 Convert 459 gallons per minute to ft³/sec.

4–3 Convert 8720 gallons per minute to ft³/sec.

4–4 Convert 84.3 gallons per minute to ft³/sec.

4–5 Convert 2.20 ft³/sec to gallons per minute.

4–6 Convert 0.064 ft³/sec to gallons per minute.

4–7 Convert 10.4 ft³/sec to gallons per minute.

4–8 Convert 0.46 ft³/sec to gallons per minute.

4–9 Convert 48.5 liters per minute to ft³/sec.

4–10 Convert 6.52 liters per second to ft³/sec.

4–11 Convert 0.23 liters per second to ft³/sec.

4–12 Convert 19.4 liters per minute to ft³/sec.

4–13 Convert 25.4 cubic centimeters per second to ft³/sec.

4–14 Convert 0.296 cubic centimeters per second to ft³/sec.

4–15 Water at 50°F is flowing at 2.50 ft³/sec. Calculate the weight flow rate and the mass flow rate.

4–16 Oil for a hydraulic system (sg = 0.90) is flowing at 0.083 ft³/sec. Calculate the weight flow rate and the mass flow rate.

4–17 A liquid refrigerant (sg = 1.08) is flowing at a weight flow rate of 6.53 lb/hr. Calculate the volume flow rate and the mass flow rate.

4–18 If a pump removes 1.65 gallons of water per minute from a tank, how long will it take to empty the tank if it contains 7425 pounds of water?

4–19 Calculate the diameter of a pipe which would carry 75.0 ft³/sec. of a liquid at an average velocity of 10.0 ft/sec.

4–20 If the velocity of a liquid is 1.65 ft/sec in a 12-inch diameter pipe, what is the velocity in a 3-inch diameter jet exiting from a nozzle attached to the pipe?

4–21 When 500 gallons per minute of water flow through a 12-inch diameter pipe which later reduces to a 6-inch diameter pipe, calculate the average velocity of flow in each pipe in ft/sec.

4–22 Water flows at 3.93 ft/sec in a 6-inch diameter pipe. Calculate the velocity of flow in a 12-inch pipe connected to it.

4–23 A 6-inch diameter pipe carries 2.50 ft³/sec of water. The pipe branches into two pipes as shown in Fig. 4–10. If the velocity in the 2-inch pipe is 40 ft/sec, what is the velocity in the 4-inch pipe?

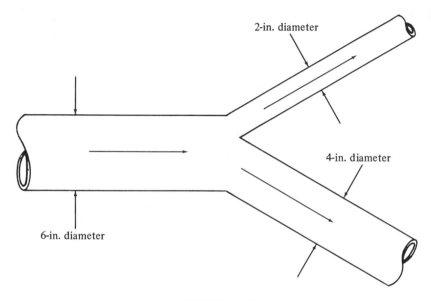

2-in. diameter

4-in. diameter

6-in. diameter

FIGURE 4–10

4–24 A standard Schedule 40 steel pipe is to be selected to carry 10 gallons per minute of water with a maximum velocity of 1.0 ft/sec. What size pipe should be used?

4–25 If water at 180°F is flowing with a velocity of 4.50 ft/sec in a standard 6-inch Schedule 40 pipe, calculate the weight flow rate in pounds per hour.

4–26 A standard 1-inch OD steel tube (0.065-inch wall thickness) is carrying 5.2 gallons per minute of oil. Calculate the velocity of flow in ft/sec.

4–27 From the list of standard steel tubing in the appendix select the smallest size which would carry 0.75 gallons per minute of oil with a maximum velocity of 1.0 ft/sec.

4–28 A standard 6-inch Schedule 40 steel pipe is carrying 95 gallons per minute of water. The pipe then branches into two standard

3-inch pipes. If the flow divides evenly between the branches, calculate the velocity of flow in all three pipes.

4–29 A shell and tube heat exchanger is made of two standard steel tubes as shown in Fig. 4–11. Each tube has a wall thickness of 0.049 inch. Calculate the required ratio of the volume flow rate in the shell to that in the tube if the average velocity of flow is to be the same in each.

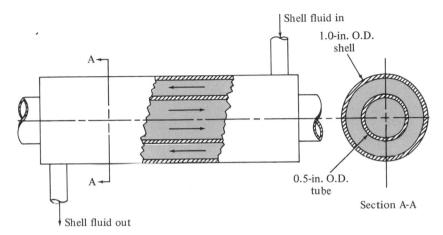

FIGURE 4–11

4–30 Gasoline (sg = 0.67) is flowing at 4.0 ft³/sec in the pipe shown in Fig. 4–12. If the pressure before the reduction is 60 psig, calculate the pressure in the 3-inch diameter pipe.

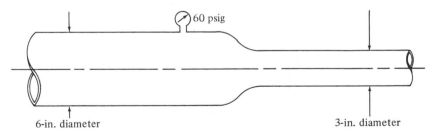

FIGURE 4–12

4–31 Water at 50°F is flowing from A to B through the pipe shown in Fig. 4–13 at the rate of 13.2 ft³/sec. If the pressure at A is 9.56 psig, calculate the pressure at B.

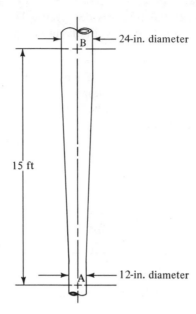

FIGURE 4–13

4–32   Calculate the volume flow rate of water at 40°F through the
system shown in Fig. 4–14.

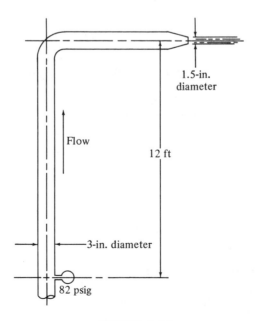

FIGURE 4–14

4–33   Calculate the required pressure in the pipe just ahead of the nozzle in Fig. 4–15 to produce a jet velocity of 75 ft/sec. The fluid is water at 180°F.

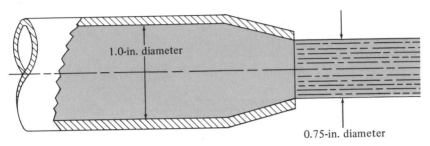

1.0-in. diameter

0.75-in. diameter

FIGURE 4–15

4–34   Kerosene having a specific weight of 50.0 lb/ft³ is flowing at 10 gallons per minute from a standard 1-inch Schedule 40 steel pipe to a standard 2-inch Schedule 40 steel pipe. Calculate the difference in pressure in the two pipes.

4–35   For the system shown in Fig. 4–16, calculate the volume flow rate of water from the nozzle and the pressure at point *A*.

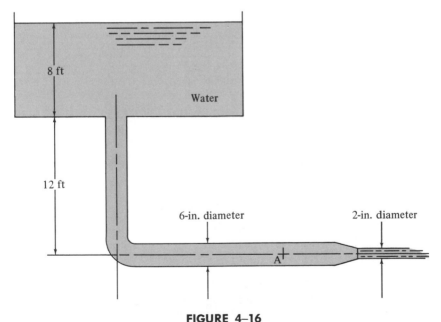

8 ft

Water

12 ft

6-in. diameter

2-in. diameter

A

FIGURE 4–16

4–36   For the system shown in Fig. 4–17, calculate the volume flow rate of oil from the nozzle and the pressures at *A* and *B*.

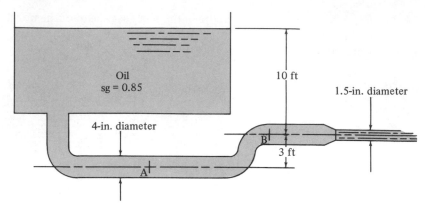

FIGURE 4–17

4–37   For the tank shown in Fig. 4–18, calculate the volume flow rate of water from the nozzle. The tank is sealed with a pressure of 20 psig above the water. The depth h is 8 feet.

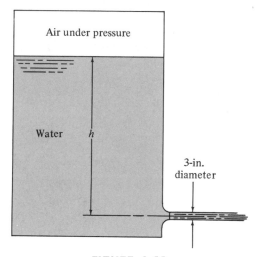

FIGURE 4–18

4–38   Calculate the pressure of the air in the sealed tank shown in Fig. 4–18 which would cause the velocity of flow to be 20 ft/sec from the nozzle. The depth h is 10 feet.

4–39   For the siphon shown in Fig. 4–19 calculate the volume flow rate of water through the nozzle and the pressures at points A and and B. The distance $X = 15$ ft and $Y = 3$ ft.

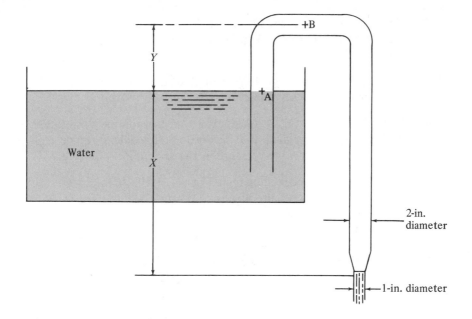

FIGURE 4–19

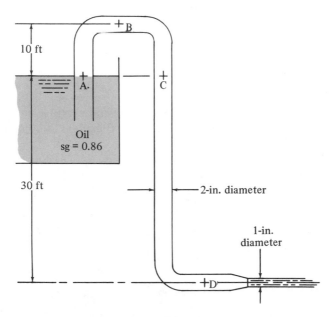

FIGURE 4–20

4–40   For the siphon shown in Fig. 4–20 calculate the volume flow
        rate of oil from the tank and the pressures at points A, B, C,
        and D.

4–41   For the siphon in Fig. 4–19, calculate the required distance X
        in order to obtain a volume flow rate of 0.25 ft³/sec.

4–42   For the siphon in Fig. 4–19, assume that the volume flow rate is
        0.20 ft³/sec. Determine the maximum allowable distance Y if
        the minimum allowable pressure in the system is -2.5 psig.

4–43   For the pipe reducer shown in Fig. 4–21 the pressure at A is
        50.0 psig and the pressure at B is 42.0 psig. Calculate the velocity
        of flow of water at point B.

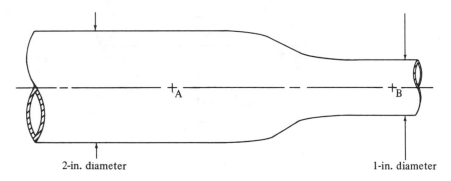

2-in. diameter   1-in. diameter

FIGURE 4–21

4–44   In the enlargement shown in Fig. 4–22 the pressure at A is 25.6
        psig and the pressure at B is 28.2 psig. Calculate the volume flow
        rate of oil (sg = 0.90).

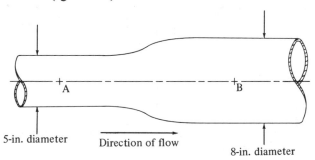

5-in. diameter   Direction of flow
                                                    8-in. diameter

FIGURE 4–22

4–45 Figure 4–23 shows a manometer being used to indicate the pressure difference between two points in a pipe system. Calculate the volume flow rate of water in the system if the manometer deflection h is 10 inches. (This arrangement is called a venturi meter, often used for flow measurement.)

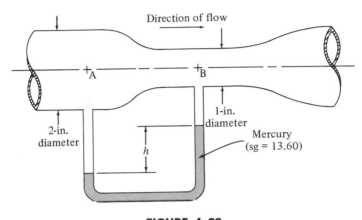

FIGURE 4–23

4–46 For the venturi meter shown in Fig. 4–23, calculate the manometer deflection h if the velocity of flow of water in the 1-inch diameter section is 30 ft/sec.

4–47 Oil with a specific weight of 55.0 lb/ft³ flows from A to B through the system shown in Fig. 4–24. Calculate the volume flow rate of oil.

4–48 The venturi meter shown in Fig. 4–25 carries water at 140°F. The specific gravity of the gage fluid in the manometer is 1.25. Calculate the volume flow rate of water.

4–49 Oil with a specific gravity of 0.90 is flowing downward through the venturi meter shown in Fig. 4–26. If the manometer deflection h is 28 inches, calculate the volume flow rate of oil.

4–50 Oil with a specific gravity of 0.90 is flowing downward through the venturi meter shown in Fig. 4–26. If the velocity of flow in the 2-inch diameter section is 10.0 ft/sec, calculate the deflection of the manometer h.

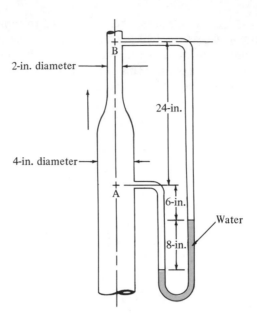

2-in. diameter

4-in. diameter

24-in.

6-in.

8-in.

Water

B

A

FIGURE 4–24

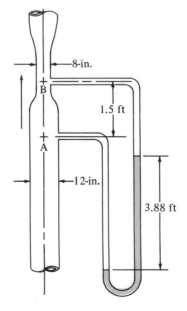

8-in.

1.5 ft

12-in.

3.88 ft

B

A

FIGURE 4–25

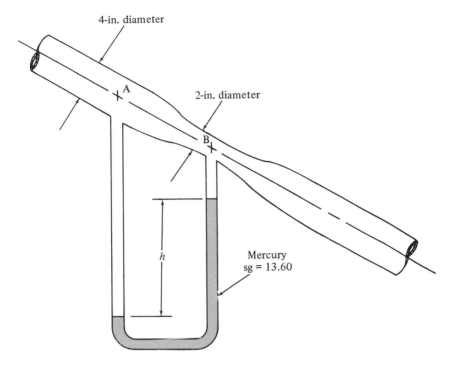

4-in. diameter

A

2-in. diameter

B

h

Mercury
sg = 13.60

FIGURE 4–26

5

General Energy Equation

Energy Losses and Additions

When Bernoulli's equation was developed in Chapter 4 there were four restrictions put on its use, namely:

1. It is valid only for incompressible fluids;
2. There can be no mechanical devices between the two sections of interest;
3. There can be no heat transferred into or out of the fluid;
4. There can be no energy lost due to friction.

Several example problems were shown in Chapter 4 where the assumption that these restrictions are satisfied would produce a negligible error in the result. However, for a flow system such as that shown in Fig. 5-1 there are definitely some energy losses and additions between the two sections of interest. For systems such as this, Bernoulli's equation is not valid.

Although the details of how to calculate the magnitude of energy losses and additions will be presented later, the general conditions under which they would occur are described below.

Mechanical Devices. With regard to their effect on a flow system, mechanical devices can be classified according to whether the device delivers energy to the fluid or the fluid delivers energy to the device.

A pump is a common example of a mechanical device which adds energy to a fluid. An electric motor or some other prime power device drives a rotating shaft in the pump. The pump then takes this kinetic energy and delivers it to the fluid resulting in increased fluid pressure

FIGURE 5–1

Typical Pipeline Installation, Showing a Pump,
Valves, Tees, and Other Fittings

Source of photo: Ingersoll-Rand Co.

and fluid flow. A wide variety of configurations are used in pump designs, some of which are shown in Fig. 5–2, 5–3, and 5–4.

Fluid motors, turbines, rotary actuators, and linear actuators are examples of devices which take energy from a fluid and deliver it in the form of work, causing the rotation of a shaft or the linear movement of a piston. Fig. 5–5 shows a photograph of a cutaway model of a typical hydraulic cylinder or linear actuator.

Many fluid motors have the same basic configurations as the pumps shown in Fig. 5–2, 5–3, and 5–4. The major difference between a pump and a fluid motor is that, when acting as a motor, the fluid drives the rotating elements of the device. The reverse is true for pumps. For some designs, such as the gear on gear type in Fig. 5–2, a pump could act as a motor by forcing a flow of fluid through the device. In other types, a change in the valve arrangement or in the configuration of the rotating elements would be required.

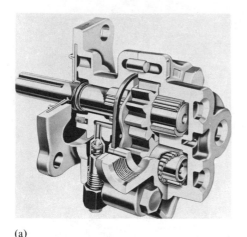

(a)

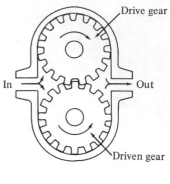

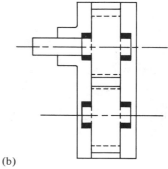

(b)

FIGURE 5–2

Gear Pump

Source of photo: Webster Electric Co., Inc.
Fluid Power Division

Source of drawing: Machine Design Magazine
Sept. 10, 1970, p. 10

Fluid Friction.   A fluid in motion offers frictional resistance to flow. Part of the energy in the system is converted into thermal energy (heat) which is dissipated through the walls of the pipe in which the fluid is flowing. The magnitude of the energy loss is dependent on the properties of the fluid, the flow velocity, the pipe size, the smoothness of the pipe wall, and the length of the pipe. Methods of calculating this frictional energy loss will be developed in later chapters.

Valves and Fittings.   Elements which control the direction or flow rate of a fluid in a system typically set up local turbulence in the fluid which causes energy to be dissipated as heat. Whenever there is a restriction, a change in flow velocity, or a change in the direction of flow, these

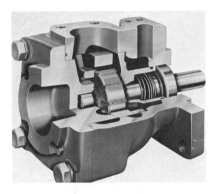

(a)

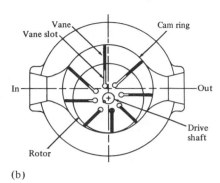

(b)

FIGURE 5-3

Vane Pump

Source of photo: Webster Electric Co., Inc.
Fluid Power Division

Source of drawing: Machine Design Magazine
Sept. 10, 1970, p. 12

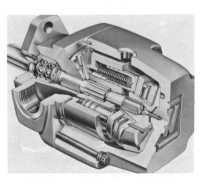

(a)

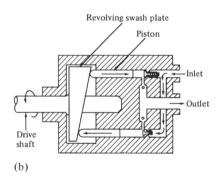

(b)

FIGURE 5-4

Piston Pump

Source of photo: Webster Electric Co., Inc.
Fluid Power Division

Source of drawing: Machine Design Magazine
Sept. 10, 1970, p. 14

FIGURE 5–5

Hydraulic Cylinder

Source: Mosier Industries
Dayton, Ohio

energy losses occur. In a large system the magnitude of losses due to valves and fittings is usually small compared with frictional losses in the pipes. Therefore, such losses are referred to as minor losses.

5–2   Nomenclature for Energy Losses and Additions

The accounting for energy losses and additions in a system will be done in the units of foot-pounds of energy per pound of fluid flowing in the system, or ft-lb/lb. This is also referred to as "feet of fluid" or "head" as described in Chapter 4. As an abbreviation for head we will use the symbol h for energy losses and additions. Specifically, the following terms will be used throughout the next several chapters.

   h_A = *Energy added* to the fluid with a mechanical device such as a pump. This is often referred to as the total head on the pump.

   h_R = *Energy removed* from the fluid by a mechanical device such as a fluid motor.

   h_L = *Energy losses* from the system due to friction in pipes or minor losses due to valves and fittings.

The magnitude of energy losses produced by many kinds of valves and fittings is directly proportional to the velocity head of the fluid. This can be expressed mathematically as

$$h_L = C_L \, (v^2/2g)$$

The term C_L is called the loss coefficient and is usually determined experimentally. Chapter 8 presents a more detailed discussion of minor losses.

5-3 General Energy Equation

The general energy equation, as it is used in this text, is an expansion of Bernoulli's equation which makes it possible to solve problems in which energy losses and additions occur. The logical interpretation of the energy equation can be seen in Fig. 5–6 which is a schematic representation of a flow system. The terms E_1 and E_2 denote the energy pos-

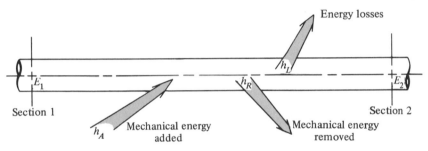

FIGURE 5–6

sessed by the fluid at sections 1 and 2, respectively. The energy additions and losses, h_A, h_R, and h_L are shown. For such a system the expression of the principle of conservation of energy is

$$E_1 + h_A - h_R - h_L = E_2 \qquad \textbf{(5-1)}$$

The energy possessed by the fluid is

$$E = \frac{p}{\gamma} + z + \frac{v^2}{2g} \qquad \textbf{(5-2)}$$

Equation (5–1) then becomes

$$\frac{p_1}{\gamma} + z_1 + \frac{v_1^2}{2g} + h_A - h_R - h_L = \frac{p_2}{\gamma} + z_2 + \frac{v_2^2}{2g} \qquad \textbf{(5-3)}$$

This is the form of the energy equation which will be used most often in this book. As with Bernoulli's Equation, each term in Eq. (5–3) represents a quantity of energy per pound of fluid flowing in the system. Typical units are ft–lb/lb, or feet of fluid flowing.

In a particular problem, it is possible that not all of the terms in the general energy equation will be required. For example, if there is no mechanical device between the sections of interest, the terms h_A and h_R will be zero and can be left out of the equation. If energy losses are so small that they can be neglected, the term h_L can be left out. If both of these conditions exist it can be seen that Eq. (5–3) reduces to Bernoulli's equation.

Programmed Example Problems

Example Problem 5–1.   Water flows from a large reservoir at the rate of 1.20 cubic feet per second through a pipe system as shown in Fig. 5–7. Calculate the total amount of energy lost from the system because of the valve, the elbows, the pipe entrance, and fluid friction.

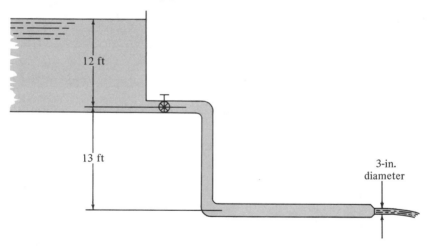

12 ft

13 ft

3-in. diameter

FIGURE 5–7

Using an approach similar to that used with Bernoulli's equation, select the two sections of interest and write the general energy equation before looking at the next panel.

The sections at which we know the most information about pressure, velocity, and elevation are the surface of the reservoir and the free stream of fluid at the exit of the pipe. Call these section 1 and section 2, respectively. Then the complete general energy equation [Eq. (5–3)] is

$$\frac{p_1}{\gamma} + z_1 + \frac{v_1^2}{2g} + h_A - h_R - h_L = \frac{p_2}{\gamma} + z_2 + \frac{v_2^2}{2g}$$

The value of some of these terms is zero. Determine which are zero and simplify the energy equation accordingly.

$p_1 = 0$ Surface of reservoir exposed to the atmosphere

$p_2 = 0$ Free stream of fluid exposed to the atmosphere

$v_1 = 0$ (approximately) Surface area of reservoir is large

$h_A = h_R = 0$ No mechanical device in the system

Then the energy equation becomes

$$z_1 - h_L = z_2 + v_2^2/2g$$

Since we are looking for the total energy lost from the system, solve this equation for h_L.

You should have

$$h_L = (z_1 - z_2) - v_2^2/2g$$

Now evaluate the terms on the right side of the equation to determine h_L in the units of ft-lb/lb.

The answer is $h_L = 15.75$ ft-lb/lb. Here is how it is done.

$$z_1 - z_2 = + 25 \text{ ft}$$
$$v_2 = Q/A_2$$

Since Q was given to be 1.20 cubic feet per second and the area of a 3-inch diameter jet is 0.0491 ft² (from the appendix),

$$v_2 = \frac{Q}{A_2} = \frac{1.20 \text{ ft}^3}{\text{sec}} \cdot \frac{1}{0.0491 \text{ ft}^2} = 24.4 \text{ ft/sec}$$

$$\frac{v_2^2}{2g} = \frac{(24.4)^2 \text{ ft}^2}{\text{sec}^2} \cdot \frac{\text{sec}^2}{(2)(32.2) \text{ ft}} = 9.25 \text{ ft}$$

Then,

$$h_L = (z_1 - z_2) - v_2^2/2g = 25 \text{ ft} - 9.25 \text{ ft}$$
$$h_L = 15.75 \text{ ft or } 15.75 \text{ ft-lb/lb}$$

Example Problem 5–2. The volume flow rate through the pump shown in Fig. 5–8 is 0.50 cubic feet per second. The fluid being pumped is oil with a specific gravity of 0.86. Calculate the energy delivered by the pump to the oil per pound of oil flowing in the system. Neglect any energy losses in the system.

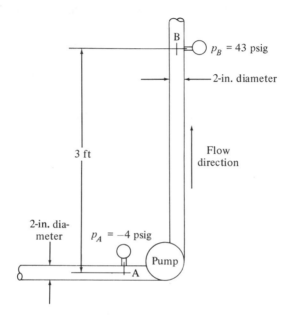

FIGURE 5–8

Using the sections where the pressure gages are located as the sections of interest, write the energy equation for the system including only the necessary terms.

You should have

$$\frac{p_A}{\gamma} + z_A + \frac{v_A^2}{2g} + h_A = \frac{p_B}{\gamma} + z_B + \frac{v_B^2}{2g}$$

Notice that the terms h_R and h_L have been left out of the general energy equation.

The objective of the problem is to calculate the energy added to the oil by the pump. Solve for h_A before looking at the next panel.

One correct solution is

$$h_A = \frac{p_B - p_A}{\gamma} + z_B - z_A + \frac{v_B{}^2 - v_A{}^2}{2g} \qquad (5\text{-}4)$$

Notice that similar terms have been grouped. This will be convenient when performing the calculations.

What is the value of the term $(v_B{}^2 - v_A{}^2)/2g$?

It is zero. Notice that the pipe size is the same at section A and at section B. The volume flow rate at each point is also the same. Then since $v = Q/A$, we can conclude that $v_A = v_B$. Now evaluate $z_B - z_A$.

You should have $z_B - z_A = 3$ ft. Now, in the remaining term, $(p_B - p_A)/\gamma$, what is the value of γ?

$\gamma = 53.6$ lb/ft^3. Remember that the specific weight of the fluid being pumped must be used.

In this case, for the oil,

$$\gamma = (sg)(\gamma_w) = (0.86)(62.4 \text{ lb/ft}^3) = 53.6 \text{ lb/ft}^3$$

Now complete the evaluation of $(p_B - p_A)/\gamma$.

The correct value is 126 feet. Since $p_B = 43$ psig and $p_A = -4$ psig,

$$\frac{p_B - p_A}{\gamma} = \frac{[(43) - (-4)]\text{lb}}{\text{in.}^2} \cdot \frac{144 \text{ in.}^2}{\text{ft}^2} \cdot \frac{\text{ft}^3}{53.6 \text{ lb}}$$

$$\frac{p_B - p_A}{\gamma} = \frac{(47)(144)}{53.6} \text{ ft} = 126 \text{ ft}$$

We can now calculate h_A from Eq. (5–4).

$$h_A = 126 \text{ ft} + 3 \text{ ft} + 0 \text{ ft} = 129 \text{ ft or } 129 \text{ ft-lb/lb}$$

That is, the pump delivers 129 ft-lb of energy to each pound of oil flowing through it.

This completes the programmed instruction.

Power is defined as the rate of doing work. In fluid mechanics it is convenient to modify this statement and consider that power is the rate at which energy is being transferred. The units for power in the English gravitational unit system are ft-lb/sec. Since it is common practice to refer to power measured in horsepower the conversion factor required is

$$1 \text{ horsepower} = 550 \text{ ft-lb/sec}$$

In the preceding example problem it was found that the pump was delivering 129 ft-lb of energy to each pound of oil as it flowed through the pump. In order to calculate the power delivered to the oil it is necessary to determine how many pounds of oil are flowing through the pump in a given amount of time. This is called the weight flow rate W which was defined in Chapter 4 and which has the units of lb/sec. Power is calculated by multiplying the energy transferred per pound of fluid by the weight flow rate. This is

$$P_A = h_A W$$

But since $W = \gamma Q$,

$$P_A = h_A \gamma Q \tag{5–5}$$

where P_A denotes power added to the fluid, γ is the specific weight of the fluid flowing through the pump and Q is the volume flow rate of the fluid.

Using the data of Example Problem 5–2, we can find the power delivered by the pump to the oil as follows.

$$P_A = h_A \gamma Q$$

But

$$h_A = 129 \text{ ft-lb/lb}$$
$$\gamma = 53.6 \text{ lb/ft}^3$$
$$Q = 0.50 \text{ ft}^3/\text{sec}$$

Then,

$$P_A = \frac{129 \text{ ft-lb}}{\text{lb}} \cdot \frac{53.6 \text{ lb}}{\text{ft}^3} \cdot \frac{0.50 \text{ ft}^3}{\text{sec}} = 3460 \frac{\text{ft-lb}}{\text{sec}}$$

Converting the result to horsepower gives

$$P_A = 3460 \text{ ft-lb/sec} \cdot \frac{1 \text{ horsepower}}{550 \text{ ft-lb/sec}} = 6.30 \text{ horsepower}$$

Mechanical Efficiency of Pumps.   The term efficiency is used to denote the ratio of the power delivered by the pump to the fluid to the power supplied to the pump. Because of energy losses due to mechanical friction in pump components, fluid friction in the pump, and excessive fluid turbulence in the pump, not all of the input power is delivered to the fluid. Then, using the symbol e_M for mechanical efficiency,

$$e_M = \frac{\text{Power delivered to fluid}}{\text{Power put into pump}} = \frac{P_A}{P_I} \qquad (5\text{–}6)$$

The value of e_M will always be less than 1.0.

Continuing with the data of Example Problem 5–2 we could calculate the power input to the pump if e_M is known. For commercially available pumps the value of e_M is published as a part of the performance data. If we assume that for the pump in this problem the efficiency is 82 percent, then

$$P_I = P_A/e_M = 6.30/0.82 = 7.69 \text{ horsepower}$$

The value of the mechanical efficiency of pumps depends not only on the design of the pump but also on the conditions under which it is operating, particularly the total head and the flow rate. For pumps used in hydraulic systems, such as those shown in Fig. 5–2, 5–3, and 5–4, efficiencies range from about 70 to 90 percent. For centrifugal pumps used primarily to transfer or circulate liquids, the efficiencies range from about 50 to 85 percent.

The following programmed example problem illustrates a possible setup for measuring pump efficiency.

Programmed Example Problem

Example Problem 5–3. For the pump test arrangement shown in Fig. 5–9 determine the mechanical efficiency of the pump if the power

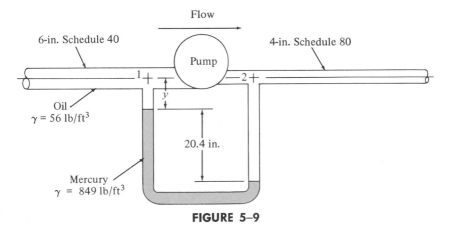

FIGURE 5–9

input is measured to be 3.85 horsepower when pumping 500 gallons per minute of oil ($\gamma = 56.0$ lb/ft³).

To begin, write the energy equation for this system.

Using the points identified as 1 and 2 in Fig. 5–9,

$$\frac{p_1}{\gamma} + z_1 + \frac{v_1^2}{2g} + h_A = \frac{p_2}{\gamma} + z_2 + \frac{v_2^2}{2g}$$

Since the power delivered by the pump to the fluid is to be found, we should now solve for h_A.

This equation is used:

$$h_A = \frac{p_2 - p_1}{\gamma} + z_2 - z_1 + \frac{v_2^2 - v_1^2}{2g} \qquad (5\text{--}7)$$

It is convenient to solve for each term individually and then combine the results. The manometer enables us to calculate $(p_2 - p_1)/\gamma$ since it measures the pressure difference. Using the procedure outlined in Chapter 3, write the manometer equation between the points 1 and 2.

Starting at point 1,

$$p_1 + \gamma_o y + \gamma_m(20.4 \text{ in.}) - \gamma_o(20.4 \text{ in.}) - \gamma_o y = p_2$$

where y is the unknown distance from point 1 to the top of the mercury column in the left leg of the manometer. The terms involving y cancel out. Also in this equation γ_o is the specific weight of the oil and γ_m is the specific weight of the mercury gage fluid.

The desired result for use in Eq. (5–7) is $(p_2 - p_1)/\gamma_o$. Solve for this now and compute the result.

The correct solution is $(p_2 - p_1)/\gamma_o = 24.0$ ft. Here is one way to do it.

$$p_2 = p_1 + \gamma_m(20.4 \text{ in.}) - \gamma_o(20.4 \text{ in.})$$

$$p_2 - p_1 = \gamma_m(20.4 \text{ in.}) - \gamma_o(20.4 \text{ in.})$$

$$\frac{p_2 - p_1}{\gamma_o} = \frac{\gamma_m(20.4 \text{ in.})}{\gamma_o} - 20.4 \text{ in.} = \left[\frac{\gamma_m}{\gamma_o} - 1\right] \cdot 20.4 \text{ in.}$$

$$= \left[\frac{849 \text{ lb/ft}^3}{56.0 \text{ lb/ft}^3} - 1\right] 20.4 \text{ in.} = (15.1 - 1)(20.4 \text{ in.})$$

$$\frac{p_2 - p_1}{\gamma_o} = (14.1)(20.4 \text{ in.}) \cdot \frac{1 \text{ ft}}{12 \text{ in.}} = 24.0 \text{ ft}$$

The next term in Eq. (5–7) is $z_2 - z_1$. What is its value?

It is zero. Both points are on the same elevation. Now find $(v_2{}^2 - v_1{}^2)/2g$.

You should have $(v_2{}^2 - v_1{}^2)/2g = 2.52$ ft.

$$Q = 500 \text{ gal/min} \cdot \frac{1 \text{ ft}^3/\text{sec}}{449 \text{ gal/min}} = 1.11 \text{ ft}^3/\text{sec}$$

Using $A_1 = 0.2006$ ft² and $A_2 = 0.07986$ ft² from the appendix,

$$v_1 = \frac{Q}{A_1} = \frac{1.11 \text{ ft}^3}{\text{sec}} \cdot \frac{1}{0.2006 \text{ ft}^2} = 5.54 \text{ ft/sec}$$

$$v_2 = \frac{Q}{A_2} = \frac{1.11 \text{ ft}^3}{\text{sec}} \cdot \frac{1}{0.07986 \text{ ft}^2} = 13.9 \text{ ft/sec}$$

$$\frac{v_2{}^2 - v_1{}^2}{2g} = \frac{(13.9)^2 - (5.54)^2}{(2)(32.2)} \frac{\text{ft}^2}{\text{sec}^2} \frac{\text{sec}^2}{\text{ft}} = 2.52 \text{ ft}$$

Now place these results into Eq. (5–7) and solve for h_A.

$$h_A = 24.0 \text{ ft} + 0 + 2.52 \text{ ft} = 26.52 \text{ ft}$$

We can now calculate the power delivered to the oil, P_A.

The result is $P_A = 3.0$ horsepower.

$$P_A = h_A \gamma Q = 26.52 \text{ ft} \cdot \frac{56.0 \text{ lb}}{\text{ft}^3} \cdot \frac{1.11 \text{ ft}^3}{\text{sec}}$$

$$P_A = 1650 \text{ ft-lb/sec} \cdot \frac{1 \text{ horsepower}}{550 \text{ ft-lb/sec}} = 3.0 \text{ horsepower}$$

The final step is to calculate e_M, the mechanical efficiency of the pump.

From Eq. (5–6) we get

$$e_M = P_A/P_I = 3.0/3.85 = 0.78$$

Expressed as a percentage, the pump is 78 percent efficient at the stated conditions.

This completes the programmed instruction.

5–5   **Power Delivered to Fluid Motors**

The energy delivered by the fluid to a mechanical device such as a fluid motor or turbine is denoted in the general energy equation by the term h_R. This is a measure of the energy in ft-lb delivered by each pound of fluid as it passes through the device. The power delivered is found by multiplying h_R by the weight flow rate W in lb/sec.

$$P_R = h_R W = h_R \gamma Q \qquad\qquad (5\text{--}8)$$

where P_R is the power delivered by the fluid to the fluid motor.

Mechanical Efficiency of Fluid Motors.   As was described for pumps, energy losses in a fluid motor are produced by mechanical and fluid friction. Therefore, not all of the power delivered to the motor is ultimately converted to power output from the device. Mechanical efficiency is then defined as

$$e_M = \frac{\text{Power output from motor}}{\text{Power delivered by fluid}} = \frac{P_O}{P_R} \qquad (5\text{--}9)$$

Here again, the value of e_M is always less than 1.0.

Programmed Example Problem

Example Problem 5–4.   Water at 50°F is flowing at a rate of 30 gal/min through the turbine shown in Fig. 5–10. The pressure at A is 100 psig

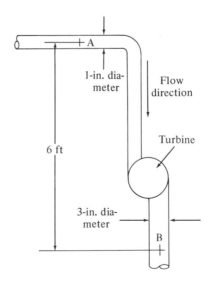

FIGURE 5–10

and the pressure at B is 25 psig. It is estimated that due to friction in the piping there is an energy loss of 10 ft-lb/lb of water flowing.

a) Calculate the horsepower delivered to the turbine by the water.
b) If the mechanical efficiency of the turbine is 85 percent, calculate the power output.

Start the solution by writing the energy equation.

Choosing points A and B as our reference points we get

$$\frac{p_A}{\gamma} + z_A + \frac{v_A^2}{2g} - h_R - h_L = \frac{p_B}{\gamma} + z_B + \frac{v_B^2}{2g}$$

The value of h_R is needed to determine the power output. Solve the energy equation for this term.

Compare this with your result.

$$h_R = \frac{p_A - p_B}{\gamma} + (z_A - z_B) + \frac{v_A^2 - v_B^2}{2g} - h_L \qquad \textbf{(5–10)}$$

Before looking at the next panel, solve for the value of each term in this equation in the unit of feet.

The correct results are:

1. $\dfrac{p_A - p_B}{\gamma} = \dfrac{(100 - 25)\ \text{lb}}{\text{in.}^2} \cdot \dfrac{\text{ft}^3}{62.4\ \text{lb}} \cdot \dfrac{144\ \text{in.}^2}{\text{ft}^2} = 173\ \text{ft}$

2. $z_A - z_B = 6\ \text{ft}$

3. $\dfrac{v_A^2 - v_B^2}{2g}$;

$$Q = 30\ \text{gal/min} \cdot \frac{1\ \text{ft}^3/\text{sec}}{449\ \text{gal/min}} = 0.0669\ \text{ft}^3/\text{sec}$$

$$v_A = \frac{Q}{A_A} = \frac{0.0669\ \text{ft}^3}{\text{sec}} \cdot \frac{1}{0.00545\ \text{ft}^2} = 12.25\ \text{ft/sec}$$

$$v_B = \frac{Q}{A_B} = \frac{0.0669\ \text{ft}^3}{\text{sec}} \cdot \frac{1}{0.0491\ \text{ft}^2} = 1.36\ \text{ft/sec}$$

$$\frac{v_A^2 - v_B^2}{2g} = \frac{(12.25)^2 - (1.36)^2}{(2)(32.2)}\ \frac{\text{ft}^2}{\text{sec}^2} \cdot \frac{\text{sec}^2}{\text{ft}} = \frac{148}{64.4}\ \text{ft} = 2.30\ \text{ft}$$

4. $h_L = 10\ \text{ft}$

Complete the solution of Eq. (5–10) for h_R now.

$$h_R = (173 + 6 + 2.3 - 10)\text{ft} = 171.3 \text{ ft}$$

To complete part (a) of the problem, calculate P_R.

The result is $P_R = 1.30$ horsepower. From Eq. (5–8),

$$P_R = h_R \gamma Q$$

$$= 171.3 \text{ ft} \cdot \frac{62.4 \text{ lb}}{\text{ft}^3} \cdot \frac{0.0669 \text{ ft}^3}{\text{sec}} = 715 \text{ ft-lb/sec}$$

$$P_R = 715 \text{ ft-lb/sec} \cdot \frac{1 \text{ horsepower}}{550 \text{ ft-lb/sec}} = 1.30 \text{ horsepower}$$

This is the power delivered to the turbine by the water. How much useful power can be expected to be put out by the turbine?

Because the efficiency of the turbine is 85 percent we get 1.10 horsepower out. From Eq. (5–9),

$$e_M = P_O/P_R$$

Then $P_O = e_M P_R = (0.85)(1.30)$ horsepower $= 1.10$ horsepower.
This completes the programmed example problem.

PRACTICE PROBLEMS

It may be necessary to refer to the appendix for data concerning the dimensions of pipes or the properties of fluids. Assume that there are no energy losses unless stated otherwise.

5–1 A horizontal pipe carries oil with a specific gravity of 0.83. If two pressure gages along the pipe read 74.6 psig and 62.2 psig, respectively, calculate the energy loss between the two gages.

5–2 Water at 40°F is flowing downward through the pipe shown in Fig. 5–11. At point A the velocity is 10 ft/sec and the pressure is 60 psig. The energy loss between points A and B is 25 ft-lb per pound of water flowing. Calculate the pressure at point B.

5–3 Find the volume flow rate of water exiting from the tank shown in Fig. 5–12. The tank is sealed with a pressure of 20 psig above the water. There is an energy loss of 6.50 ft-lb/lb as the water flows through the nozzle.

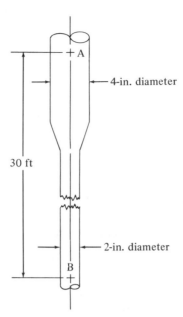

FIGURE 5–11

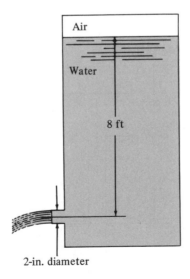

FIGURE 5–12

5–4   A long 6-inch Schedule 40 steel pipe discharges 3.0 ft³/sec of water from a reservoir into the atmosphere as shown in Fig. 5–13. Calculate the energy loss in the pipe.

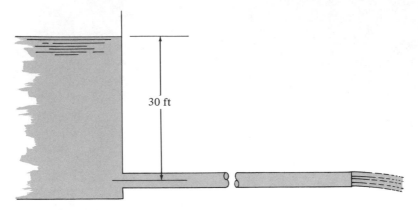

FIGURE 5-13

5-5   Figure 5-14 shows a setup to determine the energy loss due to a certain piece of apparatus. The inlet is through a 2-inch Schedule 40 pipe and the outlet is a 4-inch Schedule 40 pipe. Calculate the energy loss between points *A* and *B* if water is flowing upward at 0.20 ft³/sec. The gage fluid is mercury (sg = 13.60).

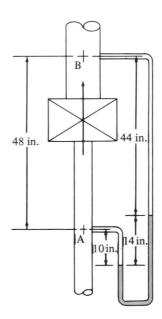

FIGURE 5-14

5-6   A test setup to determine the energy loss as water flows through a valve is shown in Fig. 5-15. Calculate the energy loss if 0.10

ft³/sec of water at 40°F is flowing. Also, calculate the loss co-efficient C_L if the energy loss is expressed as $C_L(v^2/2g)$.

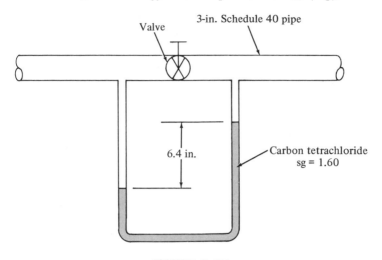

FIGURE 5–15

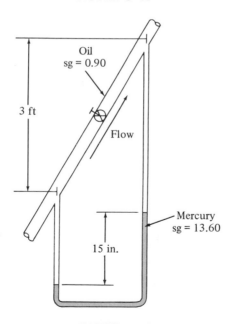

FIGURE 5–16

5–7 The setup shown in Fig. 5–16 is being used to measure the energy loss across a valve. The velocity of flow of the oil is 4.0 ft/sec.

Calculate the value of C_L if the energy loss is expressed as $C_L(v^2/2g)$.

5–8   A pump is being used to transfer water from an open tank to one having air at 75 psig above the water as shown in Fig. 5–17. If 600 gal/min are being pumped, calculate the horsepower delivered by the pump to the water. Assume that the level of the surface in each tank is the same.

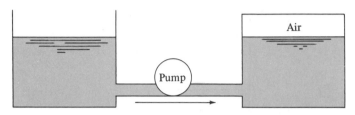

FIGURE 5–17

5–9   In problem 5–8 (Fig. 5–17), if the left hand tank was also sealed and an air pressure of 9.8 psia was above the water, calculate the pump horsepower.

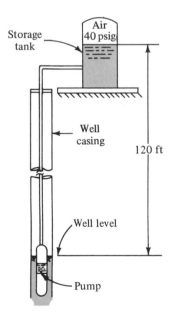

FIGURE 5–18

5–10   A commercially available sump pump is capable of delivering 2800 gallons per hour of water through a vertical lift of 20 feet. The inlet to the pump is just below the water surface and the discharge is to the atmosphere through a $1\frac{1}{4}$-inch Schedule 40 pipe. a) Calculate the power delivered by the pump to the water. b) If the pump draws $\frac{1}{2}$ horsepower, calculate its efficiency.

5–11   A submersible deep well pump delivers 745 gallons per hour of water through a 1-inch Schedule 40 pipe when operating in the system sketched in Fig. 5–18. a) Calculate the power delivered by the pump to the water. b) If the pump draws 1 horsepower calculate its efficiency.

5–12   In a pump test the suction pressure at the pump inlet is 4 psi below atmospheric pressure. The discharge pressure at a point 30 inches above the inlet is 75 psig. Both pipes are 3 inches in diameter. If the volume flow rate of water is 20 gallons per minute calculate the power delivered by the pump to the water.

5–13   The pump shown in Fig. 5–19 is delivering hydraulic oil with a specific gravity of 0.85 at a rate of 20 gallons per minute. The pressure at A is -3 psig while the pressure at B is 40 psig. The energy loss in the system is 2.5 times the velocity head in the 1-inch pipe. Calculate the horsepower delivered by the pump to the oil.

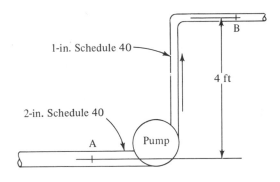

FIGURE 5–19

5–14   The pump in Fig. 5–20 delivers water from the lower to the upper reservoir at the rate of 2.0 ft³/sec. The energy loss between the suction pipe inlet and the pump is 6 ft-lb/lb and between the pump outlet and the upper reservoir is 12 ft-lb/lb. Both pipes are 6-inch Schedule 40 steel pipe. Calculate a) the pressure at the pump inlet, b) the pressure at the pump outlet, c) the total head on the pump, and d) the power delivered by the pump to the water.

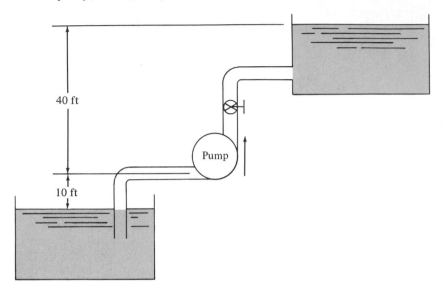

FIGURE 5–20

5–15   Repeat problem 5–14 except assume that the level of the lower reservoir is 10 feet above the pump instead of below it. All other data remains the same.

5–16   Figure 5–21 shows a pump delivering 220 gal/min of crude oil (sg = 0.85) from an underground storage drum to the first stage of a processing system. If the total energy loss in the system is 14 ft-lb/lb of oil flowing, calculate the power delivered by the pump.

5–17   Figure 5–22 shows a submersible pump being used to circulate 15 gal/min of a water based coolant (sg = 0.95) to the cutter of a milling machine. The outlet is through a ¾-inch Schedule 40 steel pipe. Assuming a total energy loss due to the piping of 9.0 ft-lb/lb, calculate the total head developed by the pump.

5–18   Figure 5–23 shows a small pump in an automatic washer discharging into a laundry sink. The washer tub is 21 inches in

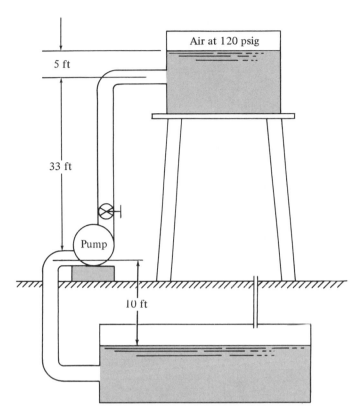

FIGURE 5-21

diameter and 10 inches deep. The average head above the pump is 15 inches as shown. The discharge hose has an inside diameter of $\frac{3}{4}$-inch. If the pump empties the tub in 1.5 minutes, calculate the average total head on the pump.

5-19 The water being pumped in the system shown in Fig. 5-24 discharges into a tank which is being weighed. It is found that 556 pounds of water are collected in 10 seconds. If the pressure at A is 2.0 psi below atmospheric pressure, calculate the horsepower delivered by the pump.

5-20 A manufacturer's rating for a gear pump states that 0.85 horsepower is required to drive the pump when it is pumping 9.1 gallons per minute of oil (sg = 0.90) with a total head of 257 feet. Calculate the mechanical efficiency of the pump.

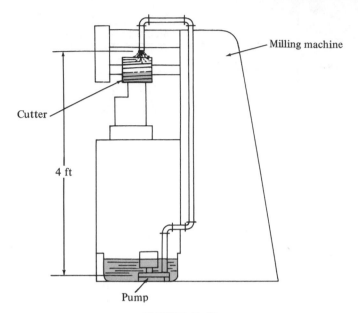

FIGURE 5–22

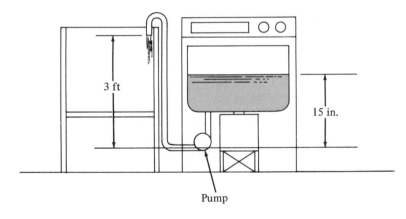

FIGURE 5–23

5–21 The specifications for an automobile fuel pump state that it should pump one pint of gasoline in 20 seconds with a suction pressure of 6 inches of mercury vacuum and a discharge pressure of 4 psig. Assuming that the pump efficiency is 60 percent, calculate the power drawn from the engine.

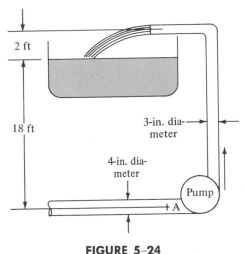

FIGURE 5–24

5–22 Figure 5–25 shows the arrangement of a circuit for a hydraulic
      system. The pump draws oil having a specific gravity of 0.90
      from a reservoir and delivers it to the hydraulic cylinder. The
      cylinder has an inside diameter of 5.0 inches and it is required
      that the piston travels 20 inches in 20 seconds while exerting a
      force of 11,000 pounds. It is estimated that there are energy
      losses of 11.5 ft-lb/lb in the suction pipe and 35.0 ft-lb/lb in
      the discharge pipe. Both pipes are ⅜-inch Schedule 80 steel pipes.

      a) Calculate the required volume flow rate through the pump.
      b) Calculate the required pressure at the cylinder.
      c) Calculate the required pressure at the outlet of the pump.
      d) Calculate the power delivered to the oil by the pump.

5–23 Calculate the horsepower delivered to the hydraulic motor in
      Fig. 5–26 if the pressure at A is 1000 psig and the pressure at
      B is 500 psig. The motor inlet is a 1-inch steel tube (0.049-inch
      wall thickness) and the outlet is a 2-inch steel tube (0.065-inch
      wall thickness). The fluid is oil (sg = 0.90) and the velocity of
      flow is 5.0 ft/sec at point B.

5–24 Water flows through the turbine shown in Fig. 5–27 at a rate of
      3400 gal/min when the pressure at A is 21.4 psig and the pressure
      at B is −5 psig. The friction energy loss between A and B is twice
      the velocity head in the 12-inch pipe. Determine the power
      delivered by the water to the turbine.

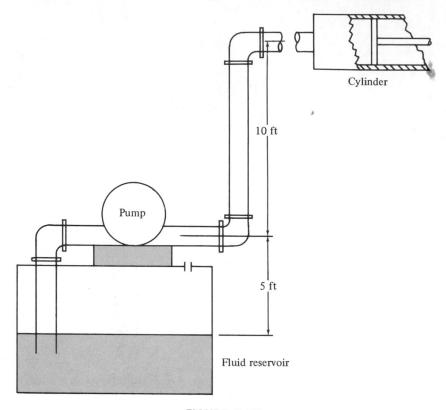

10 ft

Pump

5 ft

Cylinder

Fluid reservoir

FIGURE 5-25

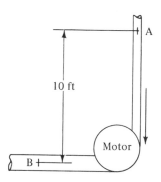

10 ft

A

Motor

B

FIGURE 5-26

5-25   Calculate the power delivered by the oil to the fluid motor shown
       in Fig. 5-28 if the volume flow rate is 7.5 ft³/sec. There is an
       energy loss of 4.5 ft-lb/lb in the piping system. If the motor has
       an efficiency of 75 percent, calculate the power output.

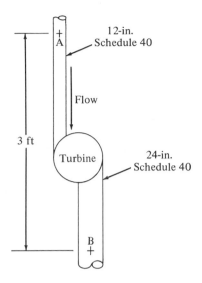

FIGURE 5–27

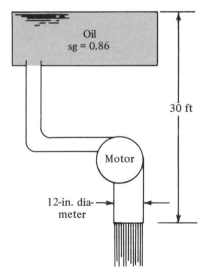

FIGURE 5–28

6

Viscosity, Laminar Flow, and Turbulent Flow

The ease with which a fluid pours is an indication of its viscosity. Cold oil has a high viscosity and pours very slowly whereas water has a relatively low viscosity and pours quite readily. Viscosity is defined as the property of a fluid which offers resistance to the relative motion of fluid molecules. The energy loss due to friction in a flowing fluid is due to the presence of its viscosity.

As a fluid moves there is developed in it a shearing stress, the magnitude of which depends on the viscosity of the fluid. Shearing stress, denoted by the Greek letter τ (tau), can be defined as the force required to slide one unit area layer of a substance over another. Thus τ is a force divided by an area and could be measured in the units of lb/ft^2.

In a fluid such as water, oil, alcohol, or other common liquids it is found that the magnitude of the shearing stress is directly proportional to the change of velocity between different positions in the fluid.

Figure 6–1 illustrates the concept of velocity change in a fluid by showing a thin layer of fluid between two surfaces, one of which is stationary while the other is moving. A fundamental condition that exists when a real fluid is in contact with a boundary surface is that the fluid has the same velocity as the boundary. Then in Fig. 6–1 the fluid in contact with the lower surface has a zero velocity and that in contact with the upper surface has the velocity v. If the distance between the two surfaces is small, then the rate of change of velocity with position y is linear. That is, it varies in a straight line manner. The velocity gradient is a measure of the velocity change and is defined as $\Delta v/\Delta y$.

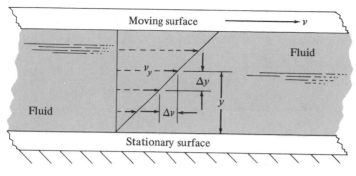

FIGURE 6–1

The fact that the shearing stress in the fluid is directly proportional to the velocity gradient can be stated mathematically as

$$\tau = \mu(\Delta v / \Delta y) \qquad (6\text{–}1)$$

where the constant of proportionality μ (the Greek letter mu) is called the dynamic viscosity of the fluid.

You can visualize the physical interpretation of Eq. (6–1) by stirring a fluid with a rod. The action of stirring causes a velocity gradient to be created in the fluid. A greater force is required to stir a cold oil having a high viscosity (a high value of μ) than that required to stir water having a low viscosity. This is an indication of the higher shearing stress in the cold oil.

The direct application of Eq. (6–1) is used in some types of viscosity measuring devices as will be explained in Section 6–4.

Units for Dynamic Viscosity. Many different unit systems are in use in this country and other countries. Because viscosity is a relatively uncommon property, the conversion factors required are not always readily available. The systems used most frequently are described here for dynamic viscosity and in the next section for kinematic viscosity. Summary tables giving many conversion factors are included in the appendix.

Because dynamic viscosity is a derived quantity, its units must be derived from Eq. (6–1). Solving for μ gives

$$\mu = \frac{\tau}{\Delta v / \Delta y} = \tau \cdot \frac{\Delta y}{\Delta v}$$

Substituting units only gives

$$\mu = \frac{\text{lb}}{\text{ft}^2} \cdot \frac{\text{ft}}{\text{ft/sec}} = \frac{\text{lb-sec}}{\text{ft}^2}$$

The table below lists the dynamic viscosity units in the three most widely used systems. The basic dimensions of force times time divided by length squared are evident in each system.

TABLE 6–1

Unit System	Dynamic Viscosity Units
English Gravitational System	$\dfrac{\text{lb-sec}}{\text{ft}^2}$ or $\dfrac{\text{slug}}{\text{ft-sec}}$
International System (SI)	$\dfrac{\text{Newton-sec}}{\text{meter}^2}$
Metric System — poise	$\dfrac{\text{dyne-sec}}{\text{cm}^2}$ or $\dfrac{\text{gram}}{\text{cm-sec}}$
Metric System — centipoise	1 centipoise $=$ 1 poise/100

6–2   **Kinematic Viscosity**

Many calculations in fluid mechanics involve the ratio of the dynamic viscosity to the density of the fluid. As a matter of convenience the kinematic viscosity ν (the Greek letter nu) is defined as

$$\nu = \mu/\rho \qquad\qquad (6\text{–}2)$$

Since μ and ρ are both properties of the fluid, ν is also a property.

Units for Kinematic Viscosity. We can derive the units for kinematic viscosity by substituting the previously developed units for μ and ρ.

$$\nu = \frac{\mu}{\rho} = \mu \cdot \frac{1}{\rho}$$
$$= \frac{\text{lb-sec}}{\text{ft}^2} \cdot \frac{\text{ft}^4}{\text{lb-sec}^2}$$
$$\nu = \text{ft}^2/\text{sec}$$

The table below lists the kinematic viscosity units in the three most widely used systems. The basic dimensions of length squared divided by time are evident.

TABLE 6-2

Unit System		Kinematic Viscosity Units
English Gravitational System		ft^2/sec
International System (SI)		meter^2/sec
Metric System	stoke	cm^2/sec
	centistoke	1 centistoke = 1 stoke/100

6-3 **Newtonian Fluids**

Any fluid which behaves in accordance with Eq. (6–1) is called a Newtonian fluid. The viscosity μ is a function only of the condition of the
fluid, particularly its temperature. The magnitude of the velocity gradient
$\Delta v/\Delta y$ has no effect on the magnitude of μ. Most common fluids such as
water, oil, gasoline, alcohol, kerosene, benzene, and glycerine are
classified as Newtonian fluids.

Conversely, a fluid which does not behave in accordance with Eq.
(6–1) is called a non-Newtonian fluid. The difference between the two
is shown in Fig. 6–2. The viscosity of the non-Newtonian fluid is dependent on velocity gradient in addition to the condition of the
fluid. Examples of this type of fluid are slurries, suspensions, gels, and
colloids. All fluids discussed in this book will be considered Newtonian
fluids unless stated otherwise.

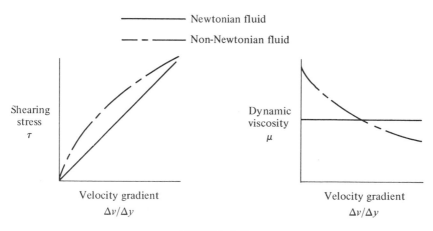

FIGURE 6-2

Procedures and equipment for measuring viscosity are numerous. Some employ fundamental principles of fluid mechanics to indicate viscosity in its basic units. Others indicate only relative values for viscosity which can be used to compare different fluids. Several common methods used for viscosity measurement are described in this section.

Rotating Drum Viscometer. The apparatus shown in Fig. 6–3 allows viscosity measurement using the definition of dynamic viscosity given by Eq. (6–3).

$$\mu = \tau/(\Delta v/\Delta y) \qquad\qquad (6\text{–}3)$$

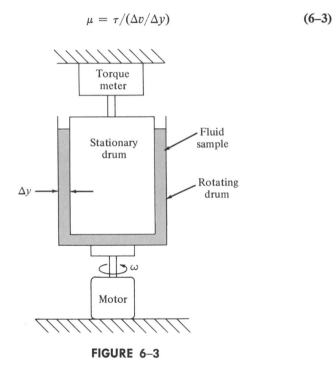

FIGURE 6–3

The outer drum is caused to rotate at a constant angular velocity, ω, while the inner drum is held stationary. Therefore, the fluid in contact with the rotating drum has a known linear velocity v while the fluid in contact with the inner drum has a zero velocity. If the thickness, Δy, of the fluid sample is known then the term $\Delta v/\Delta y$ in Eq. (6–3) can be calculated. Special consideration is given to the fluid at the bottom of the drum since its velocity is not uniform at all points. Because of the fluid viscosity, a drag force exists on the surface of the inner drum causing a torque to be developed which can be measured by a sensitive torque-meter. The magnitude of this torque is a measure of the shear stress τ in the fluid. Thus the viscosity μ can be calculated from Eq. (6–3).

Capillary Tube Viscometer. Figure 6–4 shows two reservoirs connected by a long, small diameter tube called a capillary tube. As the fluid flows through the tube with a constant velocity, some energy is lost from the

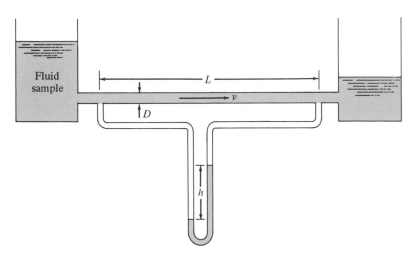

FIGURE 6–4

system causing a pressure drop which can be measured using manometers. The magnitude of the pressure drop is related to the fluid viscosity by the following equation which is developed in a later chapter of this book.

$$\mu = \frac{(p_1 - p_2)D^2}{32vL} \tag{6-4}$$

In Eq. (6–4) D is the tube inside diameter, v is the fluid velocity, and L is the length of the tube between the points 1 and 2 where the pressure is measured.

Falling Ball Viscometer. As a body falls in a fluid under the influence of gravity only, it will accelerate until the downward force (its weight) is just balanced by the buoyant force and the viscous drag force acting upward. Its velocity at that time is called the terminal velocity. The falling ball viscometer sketched in Fig. 6–5 uses this principle by causing a spherical ball to fall freely through the fluid and measuring the time required to drop a known distance. Thus the velocity can be calculated.

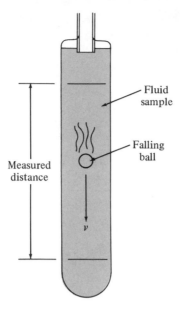

FIGURE 6–5

Figure 6–6 shows a free body diagram of the ball where w is the weight of the ball, F_b is the buoyant force, and F_d is the viscous drag force on

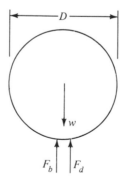

FIGURE 6–6

the ball. When the ball has reached its terminal velocity it is in equilibrium. Therefore,

$$w - F_b - F_d = 0 \tag{6-5}$$

If γ_s is the specific weight of the sphere, γ_f is the specific weight of the fluid, V is the volume of the sphere, and D is the diameter of the sphere,

$$w = \gamma_s V = \gamma_s \pi D^3 / 6 \tag{6-6}$$

$$F_b = \gamma_f V = \gamma_f \pi D^3 / 6 \tag{6-7}$$

The drag force on the sphere is

$$F_d = 3\pi\mu vD \tag{6-8}$$

Equation (6–5) then becomes

$$\gamma_s \pi D^3/6 - \gamma_f \pi D^3/6 - 3\pi\mu vD = 0$$

Solving for μ gives

$$\mu = \frac{(\gamma_s - \gamma_f)D^2}{18v} \tag{6-9}$$

Saybolt Universal Viscometer. The ease with which a fluid flows through a small diameter orifice is an indication of its viscosity. This is the principle on which the Saybolt viscometer is based. The fluid sample is placed in an apparatus similar to that sketched in Fig. 6–7. After flow is established, the time required to collect 60 cubic centimeters of the fluid is measured. The resulting time is reported as the viscosity of the fluid in Saybolt Seconds Universal (SSU or sometimes SUS). Since the measurement is not based on the basic definition of viscosity, the results are only relative. However, they do serve to compare the viscosities of different fluids. The advantage of this procedure is that it is simple and requires relatively unsophisticated equipment. Approximate conversion can be made from SSU to kinematic viscosity as shown in the appendix.

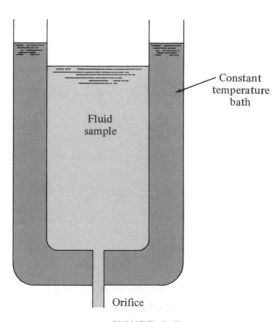

FIGURE 6–7

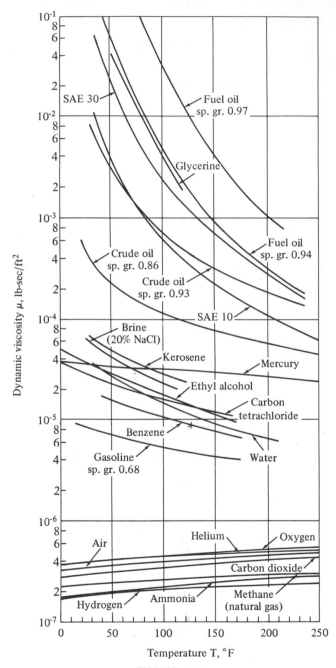

FIGURE 6–8

Source of graph: Fluid Mechanics
          R.H.F. Pao
          John Wiley & Sons, Inc., p. 173

6–5   **Variation of Viscosity with Temperature**

You are probably familiar with some examples of the variation of fluid viscosity with temperature. Engine oil is generally quite difficult to pour when it is cold, indicating that it has a high viscosity. As the temperature of the oil is increased its viscosity decreases noticeably.

All fluids exhibit this behavior to some extent. Figure 6–8 is a graph of dynamic viscosity versus temperature for many common liquids. Notice that viscosity is plotted on a logarithmic scale because of the large range of numerical values. In order to check your ability to interpret this graph, a few examples are listed below.

TABLE 6–3

Fluid	Temperature °F	Dynamic Viscosity, μ 1b-sec/ft^2
Water	70	2.0×10^{-5}
Gasoline	70	6.0×10^{-6}
SAE 30 oil	65	1.0×10^{-2}
SAE 30 oil	100	2.4×10^{-3}

Gases behave differently from liquids in that the viscosity increases as the temperature increases. Also, the amount of change is generally smaller than for liquids.

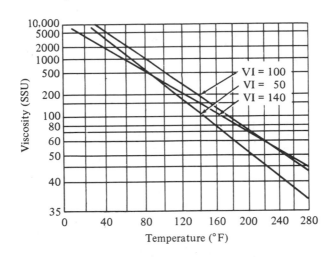

FIGURE 6–9

Typical Viscosity-Temperature Relationships

Source: Machine Design Magazine
        Sept. 10, 1970, p. 53

Viscosity Index. A measure of how greatly the viscosity of a fluid changes with temperature is given by its viscosity index, sometimes referred to as VI. This is especially important for lubricating oils and hydraulic fluids used in equipment which must operate at wide extremes of temperature. A fluid with a high viscosity index exhibits a small change in viscosity with temperature. A fluid with a low viscosity index exhibits a large change in viscosity with temperature. Typical curves for oils having viscosity indexes of 50, 100, and 140 are shown in Fig. 6–9. Viscosity index is determined by measuring the viscosity of the sample fluid at 100°F and at 210°F and comparing these values with those of certain reference fluids.

6–6   **SAE Numbers**

The Society of Automotive Engineers has developed a rating system for engine oils and gear and axle lubricants which indicates the viscosity of the oil at specific temperatures. Table 6–4 lists the specifications.

TABLE 6–4

SAE Viscosity Numbers

SAE Viscosity Number	Viscosity Range (SSU)			
	at 0°F		at 210°F	
	Min	Max	Min	Max
5W	—	6,000	—	—
10W	6,000	12,000	—	—
20W	12,000	48,000	—	—
20	—	—	45	58
30	—	—	58	70
40	—	—	70	85
50	—	—	85	110
75	—	15,000	—	—
80	15,000	100,000	—	—
90	—	—	75	120
140	—	—	120	200
250	—	—	200	—
10W-30	6,000	12,000	58	70

Source: SAE Handbook
Society of Automotive Engineers

The SAE Numbers 5W through 50 refer to crankcase oils while 75 through 250 refer to transmission and axle lubricants. Oils with the suffix W and SAE 75 and 80 are based on viscosities measured at 0°F and are meant to be used in cold temperatures. Other viscosity numbers are based on measurements at 210°F. An oil with more than one viscosity number, such as 10W-30, meets the specifications at both the 0°F and 210°F temperatures.

6–7   Laminar Flow, Turbulent Flow, and Reynolds Number

When analyzing a fluid in a flow stream it is important to be able to determine the character of the flow. Under some conditions the fluid will appear to flow in layers in a smooth and regular manner. You can observe this by opening a water faucet slowly until the flow is smooth and steady. This type of flow is called *laminar flow*. If the water faucet is opened wider allowing the velocity of flow to increase, a point would be reached when the flow is no longer smooth and regular. The water in the stream would appear to be moving in a rather chaotic manner. The flow would then be termed *turbulent flow*.

The behavior of a fluid, particularly with regard to energy losses, is quite dependent on whether the flow is laminar or turbulent as will be demonstrated in Chapter 7. For this reason it is desirable to have a means of predicting the type of flow without actually observing it. Indeed, direct observation is impossible for fluids in a rigid pipe system. It can be shown experimentally and verified analytically that the character of flow in a round pipe depends on four variables; fluid density ρ, fluid viscosity μ, pipe diameter D, and average velocity of flow v. Osborne Reynolds was the first to demonstrate that laminar or turbulent flow can be predicted if the magnitude of a dimensionless number, now called the Reynolds number N_R, is known.

$$N_R = \frac{vD\rho}{\mu} \qquad\qquad (6\text{--}10)$$

Equation (6–10) is the basic definition of the Reynolds number. If we recall that $\nu = \mu/\rho$, the Reynolds number can be written in the equivalent form

$$N_R = \frac{vD}{\nu} \qquad\qquad (6\text{--}11)$$

That the Reynolds number is dimensionless can be demonstrated by substituting units into Eq. (6–10).

$$N_R = \frac{vD\rho}{\mu} = v \cdot D \cdot \rho \cdot \frac{1}{\mu}$$

$$N_R = \frac{\text{ft}}{\text{sec}} \cdot \text{ft} \cdot \frac{\text{lb-sec}^2}{\text{ft}^4} \cdot \frac{\text{ft}^2}{\text{lb-sec}}$$

Since all units can be cancelled, N_R is dimensionless. However, it is essential that all terms in the equation be in consistent units in order to obtain the correct numerical value for N_R.

Critical Reynolds Numbers. For practical applications it is found that if the Reynolds number for the flow is less than 2000, the flow will be laminar. Also, if the Reynolds number is greater than 4000 the flow can be assumed to be turbulent. In the range of Reynolds numbers between 2000 and 4000 it is impossible to predict which type of flow exists and therefore this range is called the transition region. Typical applications involve flows which are well within the laminar flow range or well within the turbulent flow range so the existence of this region of uncertainty does not cause great difficulty. If the flow in a system is found to be in the transition region, it is the usual practice to change the flow rate or pipe diameter in order to cause the flow to be definitely laminar or turbulent. More precise analysis is then possible.

By carefully minimizing external disturbances it is possible to maintain laminar flow for Reynolds numbers as high as 50,000. However, when N_R is greater than about 4000, a minor disturbance of the flow stream will cause the flow to suddenly change from laminar to turbulent. For this reason, and since we are dealing with practical applications in this book, we will assume the following:

If $N_R < 2000$, the flow is laminar.

If $N_R > 4000$, the flow is turbulent.

In most technical situations the flow is turbulent.

6–8 **Velocity Profiles**

Unless otherwise stated, whenever the term velocity is used it is assumed to indicate the average velocity of flow found from the continuity equation, $v = Q/A$. However, in some cases it is necessary to determine the fluid velocity at a point within the flow stream. The magnitude of velocity is by no means uniform across a particular section of a pipe and the manner in which the velocity varies with position is dependent on the type of flow which exists as shown in Fig. 6–10. It was observed earlier in this chapter that the velocity of a fluid in contact with a stationary solid boundary is zero. The maximum velocity for any type of flow occurs at the center of the pipe. The reason for the different shapes of the velocity profiles is that, because of the rather chaotic motion and violent mixing of fluid molecules in turbulent flow, there is a transfer of momentum between molecules resulting in a more uniform velocity distribution than in the case of laminar flow. Since laminar flow is essentially made up of layers of fluid, the momentum transfer between molecules is less and the velocity profile becomes parabolic.

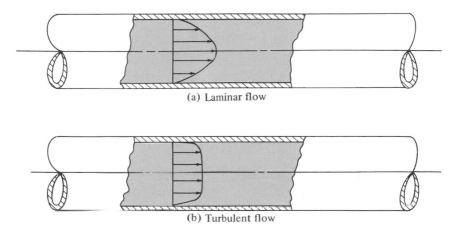

(a) Laminar flow

(b) Turbulent flow

FIGURE 6–10

It should be noted in Fig. 6–10(b) that, even though the flow as a whole is turbulent, there exists a thin layer of fluid near the pipe wall where the velocity is quite small and in which the flow is actually laminar. This is referred to as the boundary layer. The actual thickness of the boundary layer and the velocity distribution within it are very important in the analysis of heat transfer into the fluid or in determining the drag on bodies submerged in a fluid.

EXAMPLE PROBLEMS

Example Problem 6–1. Determine whether the flow is laminar or turbulent if glycerine at 77°F flows in a pipe having a 6-inch inside diameter. The average velocity of flow is 12.0 ft/sec.

Solution. The Reynolds number must be evaluated.

$$N_R = \frac{vD\rho}{\mu}$$

$$v = 12.0 \text{ ft/sec}$$

$$D = 6 \text{ in.} \times \frac{1 \text{ ft}}{12 \text{ in.}} = 0.5 \text{ ft}$$

$$\rho = 2.45 \text{ slugs/ft}^3 \text{ (from appendix)}$$

$$\mu = 1.98 \times 10^{-2} \text{ lb-sec/ft}^2 \text{ (from appendix)}$$

Then,

$$N_R = \frac{(12.0)(0.5)(2.45)}{1.98 \times 10^{-2}} = 743$$

Since $N_R = 743$ is less than 2000, the flow is laminar. Notice that each term was converted to consistent units in the English gravitational system before evaluating N_R.

Example Problem 6–2. Determine if the flow is laminar or turbulent if water at 160°F flows in a 1-inch Type L copper tube with a flow rate of 75 gallons per minute.

Solution. The Reynolds number must be evaluated.

$$N_R = \frac{vD\rho}{\mu} = \frac{vD}{\nu}$$

For a 1-inch Type L copper tube, $D = 0.08542$ ft and $A = 0.00573$ ft² from the appendix. Then,

$$v = \frac{Q}{A} = \frac{75 \text{ gal/min}}{0.00573 \text{ ft}^2} \times \frac{1 \text{ ft}^3/\text{sec}}{449 \text{ gal/min}} = 29.2 \text{ ft/sec}$$

$$\nu = 4.38 \times 10^{-6} \text{ ft}^2/\text{sec}$$

$$N_R = \frac{(29.2)(0.08542)}{4.38 \times 10^{-6}} = 5.69 \times 10^5$$

Since the Reynolds number is greater than 4000 the flow is turbulent.

Example Problem 6–3. Determine the range of average velocity of flow for which the flow would be in the transition region if SAE 10 oil at 60°F is flowing in a 2-inch Schedule 40 steel pipe. The oil has a specific gravity of 0.89.

Solution. The flow would be in the transition region if $2000 < N_R < 4000$. Then,

$$N_R = \frac{vD\rho}{\mu}$$

$$v = \frac{N_R\mu}{D\rho}$$

The values for μ, D, and ρ can be found.

$D = 0.1723$ ft (from the appendix)
$\mu = 2.10 \times 10^{-3}$ lb-sec/ft² (from Fig. 6–8)
$\rho = (\text{sg})(1.94 \text{ slugs/ft}^3) = (0.89)(1.94 \text{ slugs/ft}^3)$
$\rho = 1.73 \text{ slugs/ft}^3$

Then,

$$v = \frac{N_R(2.10 \times 10^{-3})}{(0.1723)(1.73)} = (7.05 \times 10^{-3})N_R$$

For $N_R = 2000$,

$$v = (7.05 \times 10^{-3})(2 \times 10^3) = 14.1 \text{ ft/sec}$$

For $N_R = 4000$,

$$v = (7.05 \times 10^{-3})(4 \times 10^3) = 28.2 \text{ ft/sec}$$

Therefore, if $14.1 < v < 28.2$ ft/sec the flow will be in the transition region.

PRACTICE PROBLEMS

The following problems require the use of the reference data listed below.

Fig. 6–8 Dynamic viscosity of fluids
Appendices A-C Properties of liquids
Appendices D-G Dimensions of pipe and tubing
Appendix H Conversion factors
Tables 14–2 and 14–3 Properties of air

6–1   The viscosity of a lubricating oil is given as 500 SSU. Calculate the viscosity in ft²/sec.

6–2   Using the data from Table 6–4, calculate the minimum and maximum kinematic viscosity in ft²/sec for SAE 10W-30 oil at 0°F and at 210°F.

6–3   Convert a dynamic viscosity measurement of 40 centipoises to lb-sec/ft².

6–4   Convert a kinematic viscosity measurement of 0.02 stokes to ft²/sec.

6–5   A 4-inch diameter pipe carries 0.20 ft³/sec of glycerine (sg = 1.26) at 100°F. Is the flow laminar or turbulent?

6–6   Calculate the minimum velocity of flow of water at 160°F in a 2-inch diameter pipe for which the flow is turbulent.

6–7   Calculate the maximum volume flow rate of fuel oil at 110°F at which the flow will remain laminar in a 4-inch diameter pipe. For the fuel oil, use sg = 0.895 and dynamic viscosity = 8.3 × 10⁻⁴ lb-sec/ft².

6–8   Calculate the Reynolds number for the flow of each of the following fluids in a 2-inch Schedule 40 steel pipe if the volume flow rate is 0.25 ft³/sec. (a) Water at 60°F, (b) Acetone at 77°F, (c) Castor oil at 77°F, (d) SAE 50 oil at 210°F, and (e) Air at 80°F and atmospheric pressure.

6–9   Determine the smallest Type K copper tube size which will carry 1.0 gallon per minute of the following fluids while maintaining laminar flow. (a) Water at 100°F, (b) Gasoline (sg = 0.72) at 77°F, (c) Ethyl alcohol (sg = 0.79) at 20°F, (d) SAE 50 oil at 100°F, (e) Air at 140°F and standard atmospheric pressure.

6–10 In an existing installation it is desired to carry SAE 10 oil (sg = 0.89) in a 3-inch Schedule 40 steel pipe at the rate of 225 gal/min. Efficient operation of a certain process requires that the Reynolds number of the flow be approximately 5×10^4. To what temperature must the oil be heated to accomplish this?

6–11 From the data in appendix C it can be seen that SAE 20W-40 oil and the heavy machine tool hydraulic oil have nearly the same kinematic viscosity at 210°F. However, because of their different viscosity index, the viscosities at 100°F are quite different. Calculate the Reynolds number for the flow of each oil at each temperature in a 5-inch Schedule 80 steel pipe at 10 ft/sec velocity. Are the flows laminar or turbulent?

6–12 In a falling ball viscometer, a steel ball 1/16-inch in diameter is allowed to fall freely in a heavy fuel oil having a specific gravity of 0.94. Steel weighs 490 lb/ft³. If the ball is observed to fall 10 inches in 10.4 seconds calculate the viscosity of the oil.

6–13 A capillary tube viscometer similar to that shown in Fig. 6-4 is being used to measure the viscosity of an oil having a specific gravity of 0.90. The following data apply.

Tube inside diameter = 0.10 inch = D
Length between manometer taps = 1 foot = L
Manometer fluid is mercury
Manometer deflection = 6.95 inches = h
Velocity of flow = 5.20 ft/sec = v

Determine the viscosity of the oil.

7

Energy Losses Due to Friction

Darcy's Equation

In the general energy equation,

$$\frac{p_1}{\gamma} + z_1 + \frac{v_1{}^2}{2g} + h_A - h_R - h_L = \frac{p_2}{\gamma} + z_2 + \frac{v_2{}^2}{2g}$$

the term h_L is defined as the energy loss from the system. One component of the energy loss is that due to friction in the flowing fluid which is proportional to the velocity head of the flow and to the ratio of the length to the diameter of the flow stream, for the case of flow in pipes and tubes. This is expressed mathematically as Darcy's equation,

$$h_L = f\frac{L}{D}\frac{v^2}{2g} \tag{7–1}$$

where

h_L = Energy loss due to friction (ft-lb/lb or ft)
L = Length of flow stream (ft)
D = Pipe diameter (ft)
v = Average velocity of flow (ft/sec)
f = Friction factor (dimensionless)

Darcy's equation can be used to calculate the energy loss due to friction in long straight sections of round pipe for both laminar and turbulent flow. The difference between the two is in the evaluation of the dimensionless friction factor f.

7–2   **Friction Loss in Laminar Flow**

When laminar flow exists, the fluid seems to flow as several layers one on another. Because of the viscosity of the fluid there is a shear stress created between the layers of fluid. Energy is lost from the fluid by the action of overcoming the frictional forces produced by the shear stress. Since laminar flow is so regular and orderly it is possible to derive a relationship between the energy loss and the measurable parameters of the flow system. Such a relationship is referred to as the Hagen-Poiseuille equation

$$h_L = \frac{32\ \mu L v}{\gamma D^2} \tag{7–2}$$

The parameters involved are the fluid properties of viscosity and specific weight, the geometrical features of length and pipe diameter, and the dynamics of the flow characterized by the average velocity. The Hagen-Poiseuille equation has been verified experimentally many times. It should be observed from Eq. (7–2) that the energy loss in laminar flow is independent of the condition of the pipe surface. Viscous friction losses within the fluid govern the magnitude of the energy loss.

The Hagen-Poiseuille equation is valid only for laminar flow ($N_R <$ 2000). However, it was stated earlier that Darcy's equation, Eq. (7–1), could also be used to calculate friction loss for laminar flow. If the two relationships for h_L are set equal to each other it is possible to solve for the value of the friction factor.

$$f \frac{L}{D} \frac{v^2}{2g} = \frac{32\ \mu L v}{\gamma D^2}$$

$$f = \frac{32\ \mu L v}{\gamma D^2} \cdot \frac{D2g}{L v^2} = \frac{64 \mu g}{v D \gamma}$$

But, since $\rho = \gamma/g$,

$$f = \frac{64 \mu}{v D \rho}$$

The Reynolds number is defined as $N_R = v D \rho / \mu$. Then

$$f = \frac{64}{N_R} \tag{7–3}$$

In summary, the energy loss due to friction in *laminar flow* can be calculated either from the Hagen-Poiseuille equation,

$$h_L = \frac{32\ \mu L v}{\gamma D^2}$$

or Darcy's equation,

$$h_L = f\frac{L}{D}\frac{v^2}{2g}$$

where $f = 64/N_R$.

Example Problem 7–1. Determine the energy loss if glycerine at 77°F flows 100 feet through a 6-inch diameter pipe with an average velocity of 12.0 ft/sec.

Solution. It first must be determined whether the flow is laminar or turbulent by evaluating the Reynolds number.

$$N_R = \frac{vD\rho}{\mu}$$

For glycerine at 77°F, from the appendix,

$$\rho = 2.45 \text{ slugs/ft}^3$$
$$\gamma = 78.62 \text{ lb/ft}^3$$
$$\mu = 1.98 \times 10^{-2} \text{ lb-sec/ft}^2$$

Then,

$$N_R = \frac{(12.0)(0.5)(2.45)}{1.98 \times 10^{-2}} = 743$$

Since $N_R < 2000$ the flow is laminar.
   Using Darcy's equation,

$$h_L = f\frac{L}{D}\frac{v^2}{2g}$$

$$f = \frac{64}{N_R} = \frac{64}{743} = 0.086$$

$$h_L = \frac{(0.086)(100)(12.0)^2}{(0.5)(2)(32.2)} \text{ ft}$$

$$h_L = 38.6 \text{ ft}$$

Notice that each term in each equation was expressed in the units of the English gravitational unit system. Therefore the resulting units for h_L are feet or ft-lb/lb. This means that 38.6 ft-lb of energy is lost by each pound of the glycerine as it flows along the 100 feet of pipe.

7–3   Friction Loss in Turbulent Flow

For turbulent flow of fluids in circular pipes it is most convenient to use Darcy's equation to calculate the energy loss due to friction. It is not possible to determine the friction factor f by a simple calculation as was done for laminar flow because turbulent flow does not conform to regular predictable motions. It is rather chaotic and constantly varying. For these reasons we must rely on experimental data to determine the value of f.

Tests have shown that the dimensionless number f is dependent on two other dimensionless numbers, the Reynolds number and the relative roughness of the pipe. The relative roughness is the ratio of the pipe diameter D to the average pipe wall roughness ϵ (Greek letter epsilon). Figure 7–1 illustrates pipe wall roughness as the height of the peaks of the surface irregularities. The condition of the pipe surface is very much dependent on the pipe material and the method of manufacture. For commercially available pipe and tubing, the design value of the wall roughness ϵ has been determined as shown in Table 7–1. These are only average values for new clean pipe. Some variation should be expected. After a pipe has been in service for a time the roughness could change due to the formation of deposits on the wall or due to corrosion.

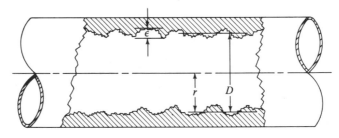

FIGURE 7–1

TABLE 7–1

Pipe Roughness — Design Values

Material	Roughness, ϵ (ft)
Glass, plastic	Smooth
Copper, brass, lead (tubing)	5×10^{-6}
Cast iron — uncoated	8×10^{-4}
Cast iron — asphalt coated	4×10^{-4}
Commercial steel or welded steel	1.5×10^{-4}
Wrought iron	1.5×10^{-4}
Riveted steel	6×10^{-3}
Concrete	4×10^{-3}

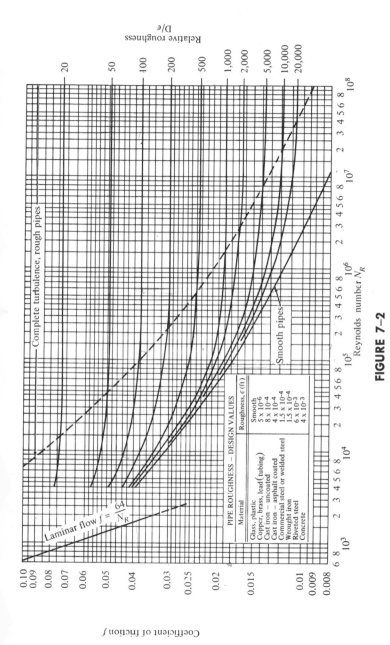

FIGURE 7-2   *Moody's Diagram*

Source of graph: Fluid Mechanics
R.H.F. Pao
John Wiley & Sons, Inc., p. 284

185

One of the most widely used methods for evaluating the friction factor employs the Moody diagram shown as Fig. 7–2. The diagram shows the friction factor f plotted versus the Reynolds number N_R with a series of parametric curves related to the relative roughness D/ϵ. These curves were generated from experimental data by L.F. Moody.

Both f and N_R are plotted on logarithmic scales because of the broad range of values encountered. At the left end of the chart, for Reynolds numbers less than 2000, the straight line shows the relationship $f = 64/N_R$ for laminar flow. For $2000 < N_R < 4000$ no curves are drawn since this is the transition zone between laminar and turbulent flow and it is not possible to predict the type of flow. Beyond $N_R = 4000$ the family of curves for different values of D/ϵ are plotted. Several important observations can be made from these curves.

1. For a given Reynolds number of flow, as the relative roughness D/ϵ is increased, the friction factor f decreases.
2. For a given relative roughness D/ϵ, the friction factor f decreases with increasing Reynolds number until the zone of complete turbulence is reached.
3. Within the zone of complete turbulence the Reynolds number has no effect on the friction factor.
4. As the relative roughness D/ϵ increases, the value of the Reynolds number at which the zone of complete turbulence begins also increases.

It should be noted that because relative roughness is defined as D/ϵ, a high relative roughness indicates a low value of ϵ, that is, a smooth pipe. In fact, the curve labeled *smooth pipes* is used for materials such as glass which have such a low roughness that D/ϵ would be an extremely large number. Some texts and references use other conventions for reporting relative roughness, such as ϵ/D, ϵ/r, or r/ϵ, where r is the pipe radius. It is felt that the convention used in this book makes calculations and interpolations easier.

Use of the Moody Diagram. The function of the Moody Diagram is to aid in determining the value of the friction factor f for turbulent flow. The value of the Reynolds number and the relative roughness must be known. Therefore, the basic data required are the pipe inside diameter, the pipe material, the flow velocity, the kind of fluid and its temperature from which the viscosity can be found. The following example problems illustrate the procedure for finding f.

Example Problem 7–2. Determine the friction factor f if water at $160°F$ is flowing at 30.0 ft/sec in a cast iron pipe having an inside diameter of one inch.

Solution. The Reynolds number must first be evaluated to determine whether the flow is laminar or turbulent.

$$N_R = \frac{vD}{\nu}$$

But D = one inch = 0.0833 ft and ν = 4.38 × 10⁻⁶ ft²/sec.

$$N_R = \frac{(30.0)(0.0833)}{4.38 \times 10^{-6}} = 5.70 \times 10^5$$

Thus the flow is turbulent. Now the relative roughness must be evaluated. From Table 7-1 we find $\epsilon = 8 \times 10^{-4}$ ft. Then

$$\frac{D}{\epsilon} = \frac{0.0833 \text{ ft}}{8 \times 10^{-4} \text{ ft}} = 1.04 \times 10^2 = 104$$

Notice that in order for D/ϵ to be a dimensionless ratio, both must be in the same units.

The final steps in the procedure are:

1. Locate the Reynolds number on the abscissa of the Moody diagram.

$$N_R = 5.70 \times 10^5$$

2. Project vertically until the curve for $D/\epsilon = 104$ is reached. Since 104 is so close to 100, that curve can be used.
3. Project horizontally to the left and read $f = 0.038$.

Example Problem 7-3. If the flow rate of water in problem 7-2 was 0.45 ft/sec with all other conditions being the same, determine the friction factor f.

Solution.

$$N_R = \frac{vD}{\nu} = \frac{(0.45)(0.0833)}{4.38 \times 10^{-6}} = 8.55 \times 10^3$$

$$\frac{D}{\epsilon} = \frac{0.0833}{8 \times 10^{-4}} = 104$$

Then from Fig. 7-2, $f = 0.044$. Notice that this is on the curved portion of the D/ϵ curve and that there is a significant increase in the friction factor over that in the preceding problem.

Example Problem 7-4. Determine the friction factor f if ethyl alcohol at 77°F is flowing at 17.5 ft/sec in a standard 1½-inch Schedule 160 steel pipe.

Solution. Evaluating the Reynolds number,

$$N_R = \frac{vD\rho}{\mu}$$

From the appendix, $\rho = 1.53$ slugs/ft^3 and $\mu = 2.29 \times 10^{-5}$ lb-sec/ft^2. Also, for a $1\frac{1}{2}$-inch, Schedule 160 pipe, $D = 0.1115$ ft. Then,

$$N_R = \frac{(17.5)(0.1115)(1.53)}{2.29 \times 10^{-5}} = 1.3 \times 10^5$$

Thus the flow is turbulent. For a steel pipe, $\epsilon = 1.5 \times 10^{-4}$ ft. Then,

$$\frac{D}{\epsilon} = \frac{0.1115 \text{ ft}}{1.5 \times 10^{-4} \text{ ft}} = 743$$

From Fig. 7–2, $f = 0.022$. Interpolation on both N_R and D/ϵ is required to determine this value and some variation should be expected. However, you should be able to read the value of the friction factor f within ± 0.0005 in this portion of the graph.

The following is a Programmed Example Problem illustrating a typical fluid piping situation. The energy loss due to friction must be calculated as a part of the solution.

Programmed Example Problem

Example Problem 7–5. In a chemical processing plant it is desired to deliver benzene at 120°F (sg = 0.86) to a point B with a pressure of 80 psig. A pump is located at a point A 70 feet below point B and the two points are connected by 780 feet of plastic pipe having an inside diameter of 2.0 inches. If the volume flow rate is 28.5 gallons per minute, calculate the required pressure at the outlet of the pump.

Write the energy equation between points A and B.

The relation is

$$\frac{p_A}{\gamma} + z_A + \frac{v_A^2}{2g} - h_L = \frac{p_B}{\gamma} + z_B + \frac{v_B^2}{2g}$$

The term h_L is required because there is an energy loss due to friction between points A and B. The point A is at the outlet of the pump and the objective of the problem is to calculate p_A. Solve algebraically for p_A now.

$$p_A = p_B + \gamma\left[z_B - z_A + \frac{v_B{}^2 - v_A{}^2}{2g} + h_L\right] \qquad (7\text{-}4)$$

What is the value of $z_B - z_A$?

$z_B - z_A = +70$ ft since point B is higher than point A. Now, what is $(v_B{}^2 - v_A{}^2)/2g$?

It is zero. Since points A and B are both in the same size pipe under steady flow, the velocities are equal.

This brings us to h_L, the energy loss due to friction between A and B. What is the first step?

The evaluation of the Reynolds number is first. The type of flow, laminar or turbulent, must be determined. Complete the calculation of the Reynolds number before looking at the next panel.

The correct value is $N_R = 9.01 \times 10^4$. Here is how it is found.

$$N_R = vD\rho/\mu$$

For a 2-inch pipe, $D = 0.1667$ ft and $A = 0.0218$ ft^2. Then

$$v = \frac{Q}{A} = \frac{28.5 \text{ gal/min}}{0.0218 \text{ ft}^2} \cdot \frac{1 \text{ ft}^3/\text{sec}}{449 \text{ gal/min}}$$
$$v = 2.92 \text{ ft/sec}$$

For benzene at 120°F with a specific gravity of 0.86,

$$\rho = (0.86)(1.94 \text{ slugs/ft}^3) = 1.67 \text{ slugs/ft}^3$$
$$\mu = 9.0 \times 10^{-6} \text{ lb-sec/ft}^2 \text{ (from Fig. 6–8)}$$

Then,

$$N_R = \frac{(2.92)(0.1667)(1.67)}{9.0 \times 10^{-6}} = 9.01 \times 10^4$$

Therefore the flow is turbulent. What relationship should be used to calculate h_L?

For turbulent flow, Darcy's equation should be used.

$$h_L = f\frac{L}{D}\frac{v^2}{2g}$$

To use the Moody diagram to find the value of f, is the value of the relative roughness D/ϵ needed?

Not in this case. Since the pipe is plastic the inner surface is smooth and we can use the curve in the Moody diagram labeled *smooth pipes*. Evaluate f now.

The result is $f = 0.018$. Now the calculation of h_L can be completed.

The correct value is $h_L = 11.15$ ft.

$$h_L = f\frac{L}{D}\frac{v^2}{2g} = \frac{(0.018)(780)(2.92)^2}{(0.1667)(2)(32.2)} \text{ ft}$$

$$h_L = 11.15 \text{ ft}$$

Going back to Eq. (7–4), p_A can be calculated.

You should have $p_A = 110.3$ psig.

$$p_A = p_B + \gamma\left[z_B - z_A + \frac{v_B{}^2 - v_A{}^2}{2g} + h_L\right]$$

$$= 80 \text{ psig} + \frac{(0.86)(62.4 \text{ lb})}{\text{ft}^3}[70 \text{ ft} + 0 + 11.15 \text{ ft}]\frac{1 \text{ ft}^2}{144 \text{ in.}^2}$$

$$= 80 \text{ psig} + 30.3 \text{ lb/in.}^2$$

$$p_A = 110.3 \text{ psig}$$

PRACTICE PROBLEMS

7–1   Crude oil is flowing vertically downward through 200 feet of 1-inch Schedule 80 steel pipe at a velocity of 2.10 ft/sec. The oil has a specific gravity of 0.86 and is at 30°F. Calculate the pressure difference between the top and bottom of the pipe.

7–2   Water at 170°F is flowing in a ½-inch Type L copper tube at a rate of 3.4 gal/min. Calculate the pressure difference between two points 150 feet apart if the tube is horizontal.

7–3   Fuel oil is flowing in a 4-inch Schedule 40 steel pipe at the maximum rate for which the flow is laminar. If the oil has a specific gravity of 0.895 and a dynamic viscosity of 8.3×10^{-4} lb-sec/ft^2, calculate the energy loss per 100 feet of pipe.

7–4   A 3-inch nominal size cast iron pipe has an actual inside diameter of 3.58 inches. The pipe is 5000 feet long and carries a lubricating oil between two points A and B such that the Reynolds number is 800. Point B is 20 feet higher than A. The oil has a specific gravity of 0.90 and a dynamic viscosity of 4×10^{-4} lb-sec/ft^2. If the pressure at A is 50 psig calculate the pressure at B.

7–5   Benzene at 140°F is flowing in a 1-inch Schedule 160 steel pipe at the rate of 0.011 ft^3/sec. The specific weight of the benzene is 54.9 lb/ft^3. Calculate the pressure difference between two points 100 feet apart if the pipe is horizontal.

7–6   As a test to determine the wall roughness of an existing pipe installation, water at 60°F is pumped through it at the rate of 60 gal/min. The pipe is standard 1½-inch commercial steel tubing having a wall thickness of 0.109 inch. Pressure gages located 100 feet apart in a horizontal run of the pipe read 150 psig and 97 psig. Determine the pipe wall roughness.

7–7   Water at 80°F flows from a storage tank through 550 feet of 6-inch nominal size cast iron pipe (actual I.D. = 6.52 inches) as shown in Fig. 7–3. Taking the energy loss due to friction into account, calculate the required head h above the pipe inlet to produce a volume flow rate of 2.50 ft^3/sec.

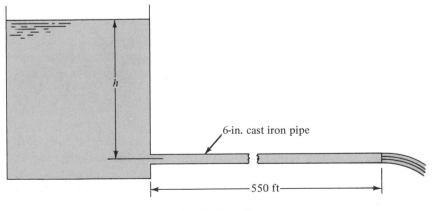

FIGURE 7–3

7–8   A water main is a concrete pressure pipe 18 inches in diameter. Calculate the pressure drop over a 1-mile length due to pipe wall friction if the pipe carries 15.0 ft³/sec of water at 50°F.

7–9   Figure 7–4 shows a portion of a fire protection system in which

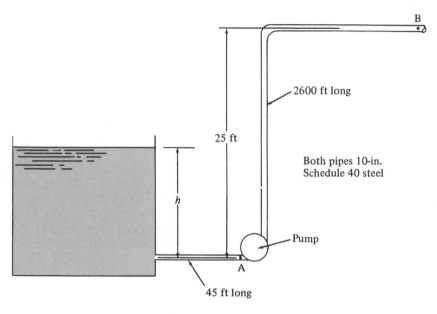

FIGURE 7–4

a pump draws water at 60°F from a reservoir and delivers it to a point *B* at the flow rate of 1500 gal/min.

a. Calculate the required height *h* of the water level in the tank in order to maintain 5.0 psig pressure at point *A*.

b. Assuming that the pressure at *A* is 5.0 psig calculate the power delivered by the pump to the water in order to maintain the pressure at point *B* at 85 psig.

Include energy losses due to friction but neglect any other energy losses.

7–10 A submersible deep well pump delivers 745 gallons per hour of water at 60°F through a one-inch Schedule 40 steel pipe when operating on the system shown in Fig. 7–5. If the total length of pipe is 140 feet, calculate the power delivered by the pump to the water.

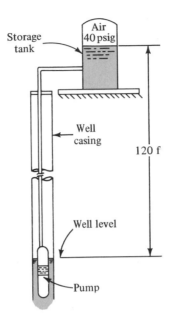

FIGURE 7–5

7–11   On a farm, water at 60°F is delivered from a pressurized storage tank to an animal watering trough through 300 feet of 1½-inch Schedule 40 steel pipe as shown in Fig. 7–6. Calculate the required air pressure above the water in the tank to produce 75 gal/min of flow.

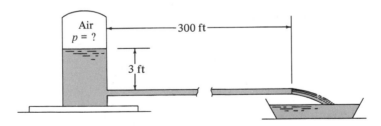

FIGURE 7–6

7–12   Figure 7–7 shows a system for delivering lawn fertilizer in liquid form. The nozzle on the end of the hose requires 20 psig of pressure to operate effectively. The hose is smooth plastic having an inside diameter of 1.0 inch. The fertilizer solution has a specific gravity of 1.10 and a dynamic viscosity of 4.0×10^{-5} lb-sec/ft². If the length of hose is 275 feet determine the power delivered by the pump to the solution and the pressure at the outlet of the pump. Neglect the energy losses on the suction side of the pump. The flow rate is 25 gal/min.

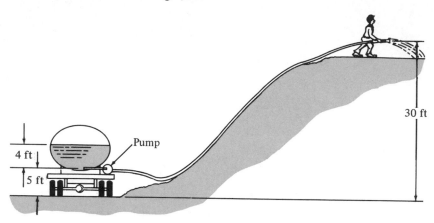

FIGURE 7-7

7-13   A pipeline transporting crude oil (sg = 0.93) at 320 gal/min is made of 6-inch Schedule 160 steel pipe. Pumping stations are spaced two miles apart. If the oil is at 60°F calculate the pressure drop between stations and the power required to maintain the same pressure at the inlet of each pump.

7-14   For the pipeline described in problem 7–13 it is proposed to heat the oil to 200°F to decrease its viscosity.

   (a) How does this affect the pump power requirement?
   (b) At what distance apart could the pumps be placed with the same pressure drop as that from problem 7–13?

8

Minor Losses

8–1   Sources of Minor Losses

In most pipe flow systems the primary energy loss is that due to pipe friction as described in the preceding chapter. Other types of losses are usually small by comparison and they are therefore referred to as minor losses. Minor losses occur whenever there is a change in the cross section of the flow path or the direction of flow, or where the flow path is obstructed as with a valve. Energy is lost under these conditions due to rather complex physical phenomena. Theoretical prediction of the magnitude of these losses is also complex and, therefore, experimental data are normally used.

The procedures developed in this chapter for analyzing minor losses were taken from many sources, some of which are listed as references at the end of the chapter. The various sets of data are presented in a form which is easily used in the analysis of pipe flow problems.

8–2   Loss Coefficient

Energy losses are proportional to the velocity head of the fluid as it flows around an elbow, through an enlargement or contraction of the flow section, or through a valve. Experimental values for energy losses are usually reported in terms of a loss coefficient, C_L, as follows.

$$h_L = C_L(v^2/2g) \qquad (8\text{–}1)$$

In Eq. (8–1), h_L is the minor loss, C_L is the loss coefficient, and v is the average velocity of flow in the pipe in the vicinity of where the minor loss occurs. In some cases, there may be more than one velocity of flow, as with enlargements or contractions. It is most important to know which velocity is to be used with each loss coefficient.

If the velocity head, $v^2/2g$, in Eq. (8–1) is expressed in the units of feet, then the energy loss, h_L, will also be in feet or foot-pounds per pound of fluid flowing. The loss coefficient is unitless, as it represents a constant of proportionality between the energy loss and the velocity head. The magnitude of the loss coefficient depends on the geometry of the device which causes the loss and sometimes on the velocity of flow. The process for determining the value of C_L and for calculating the energy loss is described in the following sections for many types of minor loss conditions.

8–3   Sudden Enlargement

As a fluid flows from a smaller pipe into a larger pipe through a sudden enlargement, its velocity abruptly decreases causing turbulence which generates an energy loss. (See Fig. 8–1.) The amount of turbulence, and therefore the amount of energy loss, is dependent on the ratio of the sizes of the two pipes.

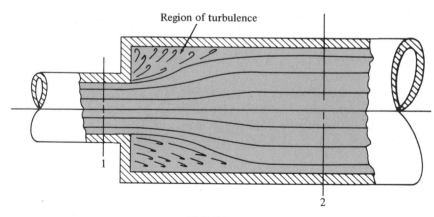

Region of turbulence

FIGURE 8–1

The minor loss is calculated from the equation,

$$h_L = C_L(v_1{}^2/2g) \tag{8–2}$$

where v_1 is the average velocity of flow in the smaller pipe ahead of the enlargement. Tests have shown that the value of the loss coefficient,

C_L, is dependent on both the ratio of the sizes of the two pipes and the magnitude of the flow velocity. This is illustrated graphically in Fig. 8–2 and in tabular form in Table 8–1.

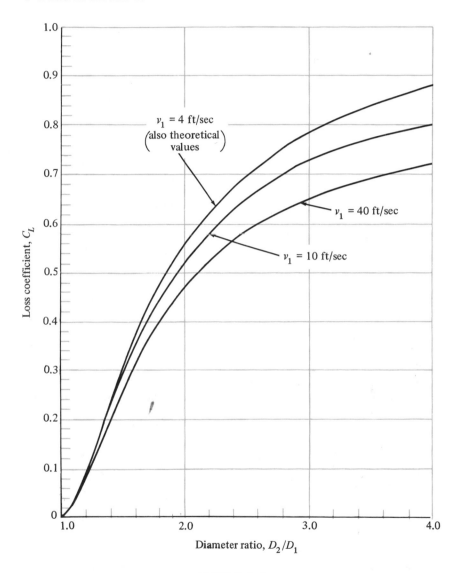

FIGURE 8–2

Sudden Enlargement

TABLE 8-1

*Loss Coefficient — Sudden Enlargement**

D₂/D₁	Theoretical C_L	Experimental C_L						
		Velocity, v_1 (ft/sec)						
D_2/D_1	C_L	2	4	10	15	20	30	40
1.0	0.0	0.0	0.0	0.0	0.0	0.0	0.0	0.0
1.2	0.10	0.11	0.10	0.09	0.09	0.09	0.09	0.08
1.4	0.24	0.26	0.25	0.23	0.22	0.22	0.21	0.20
1.6	0.37	0.40	0.38	0.35	0.34	0.33	0.32	0.32
1.8	0.48	0.51	0.48	0.45	0.43	0.42	0.41	0.40
2.0	0.56	0.60	0.56	0.52	0.51	0.50	0.48	0.47
2.5	0.71	0.74	0.70	0.65	0.63	0.62	0.60	0.58
3.0	0.79	0.83	0.78	0.73	0.70	0.69	0.67	0.65
4.0	0.88	0.92	0.87	0.80	0.78	0.76	0.74	0.72
5.0	0.92	0.96	0.91	0.84	0.82	0.80	0.77	0.75
10.0	0.98	1.00	0.96	0.89	0.86	0.84	0.82	0.80
∞	1.000	1.00	0.98	0.91	0.88	0.86	0.83	0.81

*H. W. King and E. F. Brater, *Handbook of Hydraulics* (5th ed.) (New York: McGraw-Hill Book Co., 1963).

By making some simplifying assumptions about the character of the flow stream as it expands through the sudden enlargement, it is possible to analytically predict the value of C_L from the following equation.

$$C_L = [1 - (A_1/A_2)]^2 = [1 - (D_1/D_2)^2]^2 \qquad (8\text{--}3)$$

The subscripts 1 and 2 refer to the smaller and larger section respectively as shown in Fig. 8-1. Values for C_L from this equation agree well with experimental data when the velocity v_1 is approximately 4 ft/sec. At higher velocities, the actual values of C_L are lower than the theoretical values. It is recommended that experimental values be used if the velocity of flow is known.

Example Problem 8-1. Determine the energy loss which would occur as 25 gal/min of water flows through a sudden enlargement from a 1-inch copper tube (Type K) to a 3-inch tube (Type K). See the appendix for tube dimensions.

Solution. Using the subscript 1 for the section just ahead of the enlargement and 2 for the section downstream from the enlargement,

$$D_1 = 0.995 \text{ in.} = 0.0829 \text{ ft}$$
$$A_1 = 0.00540 \text{ ft}^2$$
$$D_2 = 2.907 \text{ in.} = 0.2423 \text{ ft}$$
$$A_2 = 0.04609 \text{ ft}^2$$
$$v_1 = \frac{Q}{A_1} = \frac{25 \text{ gal/min}}{0.00540 \text{ ft}^2} \cdot \frac{1 \text{ ft}^3/\text{sec}}{449 \text{ gal/min}} = 10.3 \text{ ft/sec}$$
$$\frac{v_1^2}{2g} = \frac{(10.3)^2}{(2)(32.2)} \text{ ft} = 1.65 \text{ ft}$$

To find a value for C_L, the diameter ratio is needed.

$$D_2/D_1 = 2.907/0.995 = 2.92$$

From Fig. 8–2, $C_L = 0.72$. Then,

$$h_L = C_L(v_1^2/2g) = (0.72)(1.65) \text{ ft} = 1.19 \text{ ft}$$

This result indicates that 1.19 ft-lb of energy is dissipated from each pound of water that flows through the sudden enlargement. The problem below illustrates the calculation of the pressure difference between the two points 1 and 2.

Example Problem 8–2. Determine the difference between the pressure ahead of a sudden enlargement and the pressure downstream from the enlargement. Use the data from Example Problem 8–1.

Solution. The energy equation must be written.

$$\frac{p_1}{\gamma} + z_1 + \frac{v_1^2}{2g} - h_L = \frac{p_2}{\gamma} + z_2 + \frac{v_2^2}{2g}$$

Solving for $p_1 - p_2$ gives

$$p_1 - p_2 = \gamma[(z_2 - z_1) + (v_2^2 - v_1^2)/2g + h_L]$$

If the enlargement is horizontal, $z_2 - z_1 = 0$. Even if it were vertical, the distance between the points 1 and 2 is so small as to be considered negligible. Now, calculating the velocity in the larger pipe,

$$v_2 = \frac{Q}{A_2} = \frac{25 \text{ gal/min}}{0.04609 \text{ ft}^2} \cdot \frac{1 \text{ ft}^3/\text{sec}}{449 \text{ gal/min}} = 1.21 \text{ ft/sec}$$

Using $\gamma = 62.4$ lb/ft^3 for water, and $h_L = 1.19$ ft from the preceding problem,

$$p_1 - p_2 = \frac{62.4 \text{ lb}}{\text{ft}^3}\left[0 + \frac{(1.21)^2 - (10.3)^2}{(2)(32.2)} \text{ ft} + 1.19 \text{ ft}\right]$$

$$p_1 - p_2 = (62.4)(-1.62 + 1.19) \text{ lb/ft}^2$$

$$p_1 - p_2 = \frac{-26.8 \text{ lb}}{\text{ft}^2}\cdot\frac{1 \text{ ft}^2}{144 \text{ in.}^2} = -0.186 \text{ lb/in.}^2$$

Therefore, p_2 is 0.186 psi greater than p_1.

8–4   Exit Loss

As a fluid flows from a pipe into a large reservoir or tank as shown in Fig. 8–3, its velocity is decreased to very nearly zero. In the process, the kinetic energy which the fluid possessed in the pipe, indicated by the

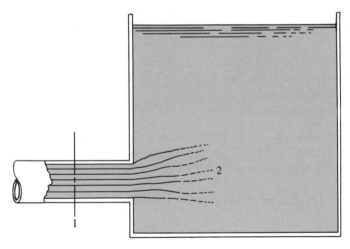

FIGURE 8–3

velocity head $v_1^2/2g$, is dissipated. Therefore, the energy loss for this condition is

$$h_L = 1.0(v_1^2/2g) \qquad (8\text{–}4)$$

which is called the exit loss.

Example Problem 8–3.   Determine the energy loss which would occur as 25 gal/min of water flows from a 1-inch copper tube (Type K) into a large tank.

Solution. Using Eq. (8–4),

$$h_L = 1.0(v_1^2/2g)$$

From the calculations in Example Problem 8–1,

$$v_1 = 10.3 \text{ ft/sec}$$
$$v_1^2/2g = 1.65 \text{ ft}$$

Then,

$$h_L = (1.0)(1.65) \text{ ft} = 1.65 \text{ ft}$$

8–5   Gradual Enlargement

If the transition from a smaller to a larger pipe can be made less abrupt than the square edge sudden enlargement, the energy loss is reduced. This is normally done by placing a conical section between the two pipes as shown in Fig. 8–4. The sloping walls of the cone tend to guide the fluid during the deceleration and expansion of the flow stream.

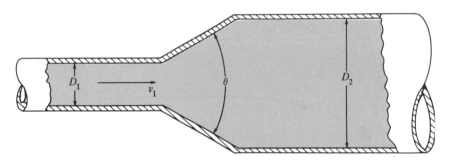

FIGURE 8–4

The energy loss for a gradual enlargement is calculated from

$$h_L = C_L(v_1^2/2g) \tag{8–5}$$

where v_1 is the velocity in the smaller pipe ahead of the enlargement. The magnitude of C_L is dependent on both the diameter ratio D_2/D_1 and the cone angle θ. Data for various values of θ and D_2/D_1 are given in Fig. 8–5 and Table 8–2.

The energy loss calculated from Eq. (8–5) does not include the loss due to friction at the walls of the transition. For relatively steep cone angles, the length of the transition is short and therefore the wall friction loss is negligible. However, as the cone angle decreases, the length of

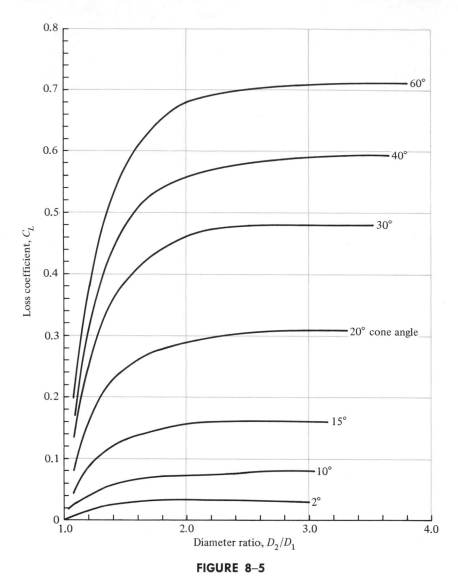

FIGURE 8–5

Gradual Enlargement

the transition increases and wall friction becomes significant. Taking both wall friction loss and the loss due to the enlargement into account, the minimum energy loss is obtained with a cone angle of about seven degrees.

TABLE 8–2

Loss Coefficient — Gradual Enlargement*

D_2/D_1	Angle of Cone, θ											
	2°	6°	10°	15°	20°	25°	30°	35°	40°	45°	50°	60°
1.1	0.01	0.01	0.03	0.05	0.10	0.13	0.16	0.18	0.19	0.20	0.21	0.23
1.2	0.02	0.02	0.04	0.09	0.16	0.21	0.25	0.29	0.31	0.33	0.35	0.37
1.4	0.02	0.03	0.06	0.12	0.23	0.30	0.36	0.41	0.44	0.47	0.50	0.53
1.6	0.03	0.04	0.07	0.14	0.26	0.35	0.42	0.47	0.51	0.54	0.57	0.61
1.8	0.03	0.04	0.07	0.15	0.28	0.37	0.44	0.50	0.54	0.58	0.61	0.65
2.0	0.03	0.04	0.07	0.16	0.29	0.38	0.46	0.52	0.56	0.60	0.63	0.68
2.5	0.03	0.04	0.08	0.16	0.30	0.39	0.48	0.54	0.58	0.62	0.65	0.70
3.0	0.03	0.04	0.08	0.16	0.31	0.40	0.48	0.55	0.59	0.63	0.66	0.71
∞	0.03	0.05	0.08	0.16	0.31	0.40	0.49	0.56	0.60	0.64	0.67	0.72

*H. W. King and E. F. Brater, *Handbook of Hydraulics* (5th ed.) (New York: McGraw-Hill Book Co., 1963).

Example Problem 8–4. Determine the energy loss which would occur as 25 gal/min of water flows from a 1-inch copper tube (Type K) into a 3-inch copper tube (Type K) through a gradual enlargement having an included cone angle of 30°.

Solution. Using data from the appendix and the results of some calculations in preceding example problems,

$$v_1 = 10.3 \text{ ft/sec}$$
$$v_1^2/2g = 1.65 \text{ ft}$$
$$D_2/D_1 = 2.907/0.995 = 2.92$$

Then from Fig. 8–5, $C_L = 0.48$, and

$$h_L = C_L(v_1^2/2g) = (0.48)(1.65) \text{ ft} = 0.79 \text{ ft}$$

Compared with the sudden enlargement described in example problem 8–1, the energy loss decreases by 34 percent when the 30° gradual enlargement is used.

8–6   **Sudden Contraction**

The energy loss due to a sudden contraction, such as that sketched in Fig. 8–6, is calculated from

$$h_L = C_L(v_2^2/2g) \tag{8–6}$$

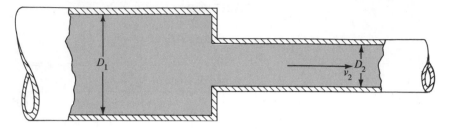

FIGURE 8–6

where v_2 is the velocity in the smaller pipe downstream from the contraction. The loss coefficient C_L is dependent on the ratio of the sizes of the two pipes and on the velocity of flow as Fig. 8–7 and Table 8–3 show.

TABLE 8–3

Loss Coefficient — Sudden Contraction*

D_1/D_2	Velocity, v_2 (ft/sec)									
	2	4	6	8	10	12	15	20	30	40
1.0	0.0	0.0	0.0	0.0	0.0	0.0	0.0	0.0	0.0	0.0
1.1	0.03	0.04	0.04	0.04	0.04	0.04	0.04	0.05	0.05	0.06
1.2	0.07	0.07	0.07	0.07	0.08	0.08	0.08	0.09	0.10	0.11
1.4	0.17	0.17	0.17	0.17	0.18	0.18	0.18	0.18	0.19	0.20
1.6	0.26	0.26	0.26	0.26	0.26	0.26	0.25	0.25	0.25	0.24
1.8	0.34	0.34	0.34	0.33	0.33	0.32	0.32	0.31	0.29	0.27
2.0	0.38	0.37	0.37	0.36	0.36	0.35	0.34	0.33	0.31	0.29
2.2	0.40	0.40	0.39	0.39	0.38	0.37	0.37	0.35	0.33	0.30
2.5	0.42	0.42	0.41	0.40	0.40	0.39	0.38	0.37	0.34	0.31
3.0	0.44	0.44	0.43	0.42	0.42	0.41	0.40	0.39	0.36	0.33
4.0	0.47	0.46	0.45	0.45	0.44	0.43	0.42	0.41	0.37	0.34
5.0	0.48	0.47	0.47	0.46	0.45	0.45	0.44	0.42	0.38	0.35
10.0	0.49	0.48	0.48	0.47	0.46	0.46	0.45	0.43	0.40	0.36
∞	0.49	0.48	0.48	0.47	0.47	0.46	0.45	0.44	0.41	0.38

*H. W. King and E. F. Brater, *Handbook of Hydraulics* (5th ed.) (New York: McGraw-Hill Book Co., 1963).

The mechanism by which energy is lost due to a sudden contraction is quite complex. Figure 8–8 illustrates what happens as the flow stream converges. The lines in the figure represent the paths of various

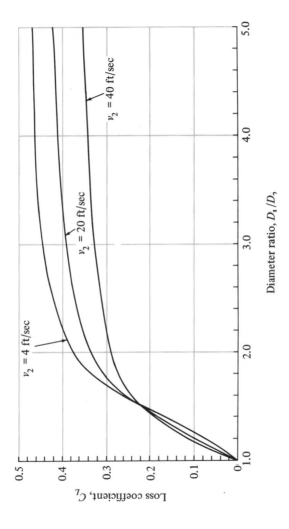

FIGURE 8-7

Sudden Contraction

parts of the flow stream called streamlines. As the streamlines approach the contraction, they assume a curved path and the total stream continues to neck down for some distance beyond the contraction. Thus the minimum cross section of the flow is smaller than that of the smaller pipe. The section where this minimum flow area occurs is called the vena contracta. Beyond the vena contracta, the flow stream must decelerate and expand to again fill the pipe. The turbulence caused by the contraction and the subsequent expansion generates the energy loss.

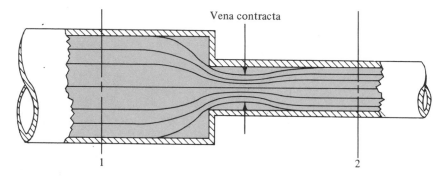

Vena contracta

FIGURE 8–8

Example Problem 8–5. Determine the energy loss which would occur as 25 gal/min of water flows from a 3-inch copper tube (Type K) into a 1-inch copper tube (Type K) through a sudden contraction.

Solution.   From Eq. (8–6),

$$h_L = C_L(v_2^2/2g)$$

For the copper tube, $D_1 = 2.907$ in., $D_2 = 0.995$ in., and $A_2 = 0.0054$ ft². Then,

$$D_1/D_2 = 2.907/0.995 = 2.92$$
$$v_2 = \frac{Q}{A_2} = \frac{25 \text{ gal/min}}{0.0054 \text{ ft}^2} \cdot \frac{1 \text{ ft}^3/\text{sec}}{449 \text{ gal/min}} = 10.3 \text{ ft/sec}$$
$$v_2^2/2g = 1.65 \text{ ft}$$

From Fig. 8–7 we can find $C_L = 0.42$. Then,

$$h_L = C_L(v_2^2/2g) = (0.42)(1.65) \text{ ft} = 0.69 \text{ ft}$$

The energy loss in a contraction can be decreased substantially by making it more gradual. However, data are lacking for numerical values of the loss coefficient. Some sources suggest using $C_L = 0.05$ for a smooth gradual contraction.

8–7   **Entrance Loss**

A special case of a contraction occurs when a fluid flows from a relatively large reservoir or tank into a pipe. The fluid must accelerate
from a negligible velocity to the flow velocity in the pipe. The ease with
which the acceleration is accomplished determines the amount of energy
loss and therefore, the value of the entrance loss coefficient is dependent
on the geometry of the entrance. Figure 8–9 shows four different configurations and the suggested value of C_L for each. The streamlines

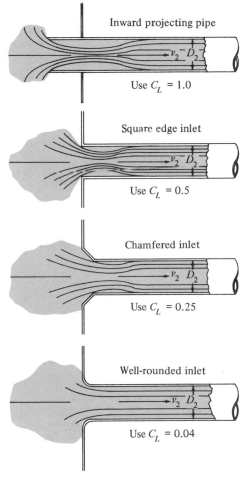

FIGURE 8–9

Entrance Loss Coefficients

illustrate the flow of fluid into the pipe and show that the turbulence associated with the formation of a vena contracta in the tube is a major cause of the energy loss. For the well-rounded entrance, no vena contracta is formed and the energy loss is quite small, due to surface friction only. Using the loss coefficient from Fig. 8–9, the energy loss at an entrance can be calculated from

$$h_L = C_L(v_2^2/2g) \tag{8–7}$$

where v_2 is the velocity of flow in the pipe.

Example Problem 8–6. Determine the energy loss which would occur as 25 gal/min of water flows from a reservoir into a 1-inch copper tube (Type K), (a) through an inward projecting tube, and (b) through a well-rounded inlet.

Solution.   Part (a)

For the tube, $D_2 = 0.995$ in. and $A_2 = 0.0054$ ft². Then,

$$v_2 = Q/A_2 = 10.3 \text{ ft/sec}$$
$$v_2^2/2g = 1.65 \text{ ft}$$

For an inward projecting entrance, $C_L = 1.0$. Then,

$$h_L = (1.0)(1.65) \text{ ft} = 1.65 \text{ ft}$$

Part (b)

For a well-rounded inlet, $C_L = 0.04$. Then,

$$h_L = (0.04)(1.65) \text{ ft} = 0.066 \text{ ft}$$

8–8   **Equivalent Length Technique**

Energy losses are calculated in terms of the dimensionless loss coefficient C_L for changes in the flow cross section as discussed in the preceding sections. The value of C_L is unaffected by the physical size of such devices. This is not true for valves and fittings such as elbows. For these types of devices the energy loss can be calculated using the equivalent length technique from the equation shown below.

$$h_L = f\frac{L_e}{D}\frac{v^2}{2g} \tag{8–8}$$

There is an obvious similarity between this equation and Darcy's equation used to calculate friction loss in pipes. The term L_e in Eq. (8–8) is the equivalent length and is defined as the length of straight pipe which would have the same total energy loss as the valve or fitting.

For many commercially available valves and fittings the ratio L_e/D is approximately constant for a particular type of device regardless of size. The resistance to flow then varies with the value of the friction factor f.

A similarity also exists between Eq. (8–8) and Eq. (8–1) in which the loss coefficient C_L is used to calculate energy losses. It is sometimes convenient to convert an equivalent length to a loss coefficient and vice versa. This can be done simply. The two equations for energy loss are:

$$h_L = C_L(v^2/2g) \qquad\qquad (8\text{--}1)$$
$$h_L = f(L_e/D)(v^2/2g) \qquad\qquad (8\text{--}8)$$

If in each of these equations, v is the velocity in the pipe to which the valve or fitting is attached, then

$$C_L = f(L_e/D) \qquad\qquad (8\text{--}9)$$

Valves and Fittings. Manufacturers differ widely in the methods used to report energy loss data for valves and fittings. Some report the equivalent length ratio L_e/D, where D is the inside diameter of the standard pipe for which the valve was designed. Table 8–4 lists representative

TABLE 8–4

Resistance in Valves and Fittings

Expressed as Equivalent Length in Pipe Diameters, L_e / D

Type	Equivalent Length in Pipe Diameters, L_e/D
Globe valve — fully open	340
Angle valve — fully open	145
Gate valve — fully open	13
— $\frac{3}{4}$ open	35
— $\frac{1}{2}$ open	160
— $\frac{1}{4}$ open	900
Check valve — swing type	135
Check valve — ball type	150
Butterfly valve — fully open	40
90° standard elbow	30
90° long radius elbow	20
90° street elbow	50
45° standard elbow	16
45° street elbow	26
Close return bend	50
Standard tee — with flow through run	20
— with flow through branch	60

Source: The Crane Co.

values of L_e/D for typical plumbing and process control valves and fittings. Equation (8–8) would be used to calculate the energy loss. Other manufacturers report the loss coefficient C_L (sometimes called K) for a particular type and size of device. Equation (8–1) must then be used.

The physical appearance of typical valves and fittings such as those listed in Table 8–4 are shown in Figs. 8–10 through 8–17. The magnitude of the equivalent length ratio L_e/D, and therefore of the energy loss, is dependent on the complexity of the fluid path through the device.

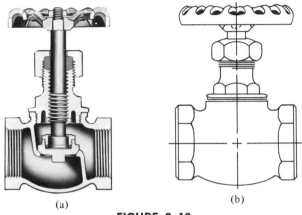

(a)   (b)

FIGURE 8–10

Globe Valve

Source: The Crane Co.

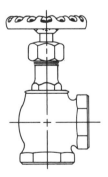

FIGURE 8–11

Angle Valve

Source: The Crane Co.

The internal construction of the angle valve is similar to the globe valve except that the fluid flows directly through the seat and then turns 90° as it leaves the valve. In the gate valve, the gate is lifted out of the flow

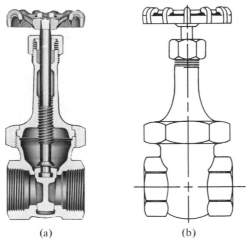

(a)   (b)

FIGURE 8–12

Gate Valve

Source: The Crane Co.

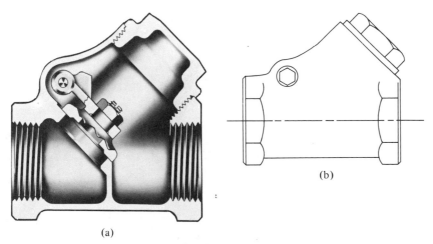

(b)

(a)

FIGURE 8–13

Check Valve — Swing Type

Source: The Crane Co.

stream as it is opened. When fully open, there is only a very minor obstruction. The function of the check valve is to allow flow in only one direction.

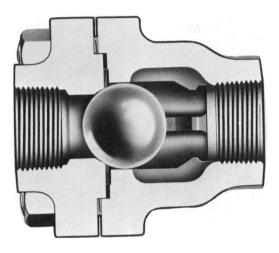

FIGURE 8–14

Check Valve — Ball Type

Source: The Crane Co.

FIGURE 8–15

Butterfly Valve

Source: The Crane Co.

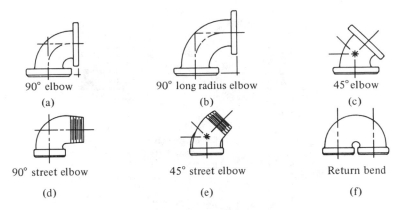

90° elbow   90° long radius elbow   45° elbow

(a)   (b)   (c)

90° street elbow   45° street elbow   Return bend

(d)   (e)   (f)

FIGURE 8–16

Pipe Elbows

Source: The Crane Co.

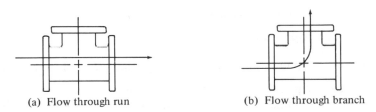

(a) Flow through run   (b) Flow through branch

FIGURE 8–17

Standard Tees

Source: The Crane Co.

In some cases, particularly with regard to control valves in fluid power systems, energy loss as such is not reported. Instead, the magnitude of the pressure drop as the fluid flows through the valve at a certain flow rate is reported. An example of a directional control valve is shown in a cutaway view in Fig. 8–18. Pressure drop data for this type of valve may be reported as follows:

Pressure Drop (psi)	10	20	30	40	50
Flow Rate (gal/min)	7.6	10.9	13.6	16.4	18.1

Source: Racine Hydraulics

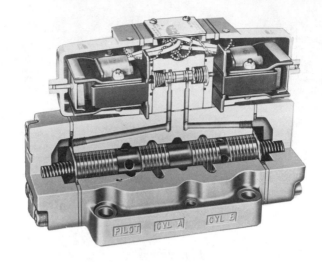

FIGURE 8–18

Directional Control Valve

Source: Racine Hydraulics

The pressure drop increases approximately as the square of the flow rate.

Bends. It is frequently more convenient to bend a pipe or tube rather than install a commercially made elbow. The resistance to flow of a bend is dependent on the ratio of the bend radius r to the pipe inside diameter D. Figure 8–19 shows that the minimum resistance occurs when the ratio r/D is approximately three. The resistance is given in terms of the equivalent length ratio L_e/D, and, therefore, Eq. (8–8) must be used to calculate the energy loss. The resistance shown in Fig. 8–19 includes both the bend resistance and the resistance due to the length of the pipe in the bend.

Example Problem 8–7. Determine the equivalent length in feet of pipe of a fully open globe valve placed in a 6-inch Schedule 40 pipe.

Solution. From Table 8–4 it is found that the equivalent length ratio L_e/D for a fully open globe valve is 340. The actual inside diameter of a 6-inch Schedule 40 pipe is 0.5054 foot. Then,

$$L_e = (L_e/D)(D) = (340)(0.5054) \text{ ft} = 172 \text{ ft}$$

Example Problem 8–8. Calculate the pressure drop across a fully open globe valve placed in a 4-inch Schedule 40 steel pipe carrying 400 gal/min of SAE 10W oil at 100°F.

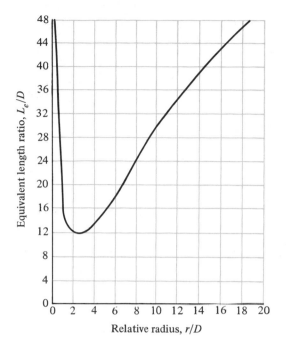

FIGURE 8–19

Resistance Due to 90° Pipe Bends*

Solution. A sketch of the installation is shown in Fig. 8–20. In order to determine the pressure drop, the energy equation should be written between the points 1 and 2.

$$\frac{p_1}{\gamma} + z_1 + \frac{v_1{}^2}{2g} - h_L = \frac{p_2}{\gamma} + z_2 + \frac{v_2{}^2}{2g}$$

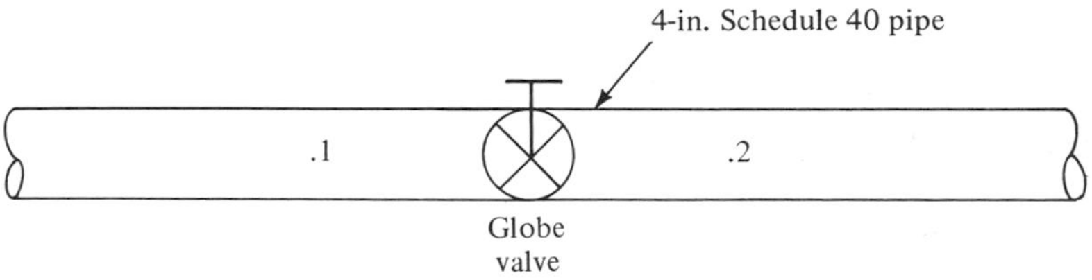

4-in. Schedule 40 pipe

.1   .2

Globe
valve

FIGURE 8–20

---

*Beij, K. H. "Pressure Losses for Fluid Flow in 90 Degree Pipe Bends." *Journal of Research of the National Bureau of Standards*, XXI (July, 1938).

The energy loss h_L is the minor loss due to the valve only. The pressure drop is the difference between p_1 and p_2. Solving the energy equation for this difference gives

$$p_1 - p_2 = \gamma \left[z_2 - z_1 + \frac{v_2{}^2 - v_1{}^2}{2g} + h_L \right]$$

But $z_1 = z_2$ and $v_1 = v_2$. Then,

$$p_1 - p_2 = \gamma h_L$$

Equation (8–8) is used to determine h_L.

$$h_L = f \frac{L_e}{D} \frac{v^2}{2g}$$

For the globe valve, $L_e/D = 340$. The velocity v is the average velocity of flow in the 4-inch pipe. For the pipe, $D = 0.3355$ ft and $A = 0.0884$ ft². Then,

$$v = \frac{Q}{A} = \frac{400 \text{ gal/min}}{0.0884 \text{ ft}^2} \cdot \frac{1 \text{ ft}^3/\text{sec}}{449 \text{ gal/min}} = 10.1 \text{ ft/sec}$$

In order to evaluate the friction factor f, the Reynolds number and the relative roughness must be determined.

$$N_R = \frac{vD}{\nu} = \frac{(10.1)(0.3355)}{4.41 \times 10^{-4}} = 7.68 \times 10^3$$

$$\frac{D}{\epsilon} = \frac{0.3355}{1.5 \times 10^{-4}} = 2240$$

Then from the Moody diagram, Fig. 7–2, $f = 0.033$. Now, evaluating h_L,

$$h_L = f \frac{L_e}{D} \frac{v^2}{2g} = (0.033)(340) \frac{(10.1)^2}{(2)(32.2)} \text{ ft}$$
$$h_L = 17.8 \text{ ft}$$

For SAE 10W oil at 100°F, $\gamma = (0.870)(62.4 \text{ lb/ft}^3)$. Then,

$$p_1 - p_2 = \gamma h_L = \frac{(0.870)(62.4) \text{ lb}}{\text{ft}^3} \cdot 17.8 \text{ ft} \cdot \frac{1 \text{ ft}^2}{144 \text{ in.}^2}$$
$$p_1 - p_2 = 6.7 \text{ psi}$$

Therefore, the pressure in the oil drops by 6.7 psi as it flows through the valve. Also, an energy loss of 17.8 ft-lb is dissipated as heat from each pound of oil that flows through the valve.

Summary — Minor Losses

Sudden Enlargement: $h_L = C_L \, (v_1{}^2/2g)$

Theoretical $C_L = (1 - A_1/A_2)^2 = (1 - D_1{}^2/D_2{}^2)^2$

Experimental C_L (See Table 8–1 or Fig. 8–2.)

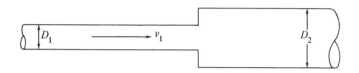

Exit Loss: $h_L = 1.0 \, (v_1{}^2/2g)$

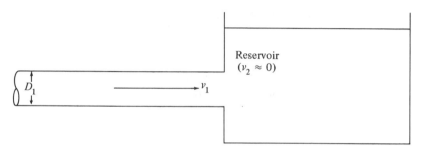

Gradual Enlargement: $h_L = C_L \, (v_1{}^2/2g)$

Experimental C_L (See Table 8–2 or Fig. 8–5.)

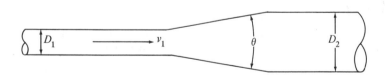

Sudden Contraction: $h_L = C_L \, (v_2{}^2/2g)$

Experimental C_L (See Table 8–3 or Fig. 8–7.)

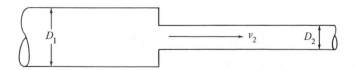

Entrance Loss: $h_L = C_L (v_2^2/2g)$

Experimental C_L (See Fig. 8–9.)

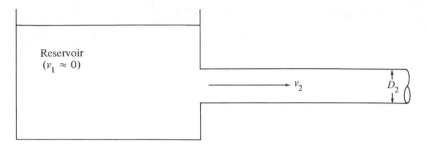

Valves, Fittings, and Bends: $h_L = f(L_e/D)(v^2/2g)$

$\quad L_e/D =$ Equivalent Length Ratio (See Table 8–4.)

$\quad\quad L_e =$ Equivalent length

$\quad\quad D =$ Pipe diameter

$\quad\quad v =$ Velocity of flow in pipe

$\quad\quad f =$ Friction factor from Fig. 7–2

References

Beij, K. H. "Pressure Losses for Fluid Flow in 90 Degree Pipe Bends," *Journal of Research of the National Bureau of Standards*, XXI (July, 1938).

Dodge, L. "How to Compute and Combine Fluid Flow Resistances in Components." *Hydraulics and Pneumatics*, September, 1968, 118–21.

Flow of Fluids through Valves, Fittings, and Pipe. Technical Paper No. 410, Crane Co., Chicago, 1969.

King, H. W., and E. F. Brater. *Handbook of Hydraulics*. New York: McGraw-Hill Book Co., (5th ed.), 1963.

King, R. C., and Sabin Crocker. *Piping Handbook*. New York: McGraw-Hill Book Co., (5th ed.), 1967.

Pipe Friction Manual. The Hydraulic Institute, New York, (3rd ed.), 1961.

Simpson, L. L. "Sizing Piping for Process Plants." *Chemical Engineering*, (June 17, 1968), 193–214.

PRACTICE PROBLEMS

8–1 Determine the energy loss due to a sudden enlargement from a 2-inch pipe to a 4-inch pipe when the velocity of flow is 10 ft/sec in the smaller pipe.

8–2 Determine the energy loss due to a sudden enlargement from a standard 1-inch Schedule 80 pipe to a $3\frac{1}{2}$-inch Schedule 80 pipe when the rate of flow is 0.10 ft³/sec.

8–3 Determine the pressure difference between two points on either side of a sudden enlargement from a tube with a 2-inch inside diameter to one with a 6-inch inside diameter when the velocity of flow of water is 4 ft/sec in the smaller tube.

8–4 Determine the pressure difference for the conditions of problem 8–3 if the enlargement was gradual with a cone angle of 15° instead of sudden.

8–5 Determine the energy loss due to a gradual enlargement from a 1-inch pipe to a 3-inch pipe when the velocity of flow is 10 ft/sec in the smaller pipe and the cone angle of the enlargement is 20°.

8–6 Determine the energy loss for the conditions of problem 8–5 if the cone angle is increased to 60°.

8–7 Determine the energy loss when 1.50 ft³/sec of water flows from a 6-inch standard Schedule 40 pipe into a large reservoir.

8–8 Determine the energy loss when oil having a specific gravity of 0.87 flows from a 4-inch pipe to a 2-inch pipe through a sudden contraction if the velocity of flow in the larger pipe is 4.0 ft/sec.

8–9 For the conditions of problem 8–8, if the pressure before the contraction were 80 psig, calculate the pressure in the smaller pipe.

8–10 Is this statement true or false? For a sudden contraction with a diameter ratio of 3.0, the energy loss decreases as the velocity of flow increases.

8–11 Determine the energy loss which will occur if water flows from a reservoir into a pipe with a velocity of 10 ft/sec if the configuration of the entrance is (a) an inward-projecting pipe, (b) a square-edged inlet, (c) a chamfered inlet, or (d) a well-rounded inlet.

8–12 Determine the equivalent length in feet of pipe of a fully open globe valve placed in a 10-inch Schedule 40 pipe. Repeat the calculation for a gate valve.

8–13   Calculate the loss coefficient C_L for a ball-type check valve placed in a 2-inch Schedule 40 steel pipe if water at 100°F is flowing with a velocity of 10 ft/sec.

8–14   Calculate the pressure difference across a fully open angle valve placed in a 5-inch Schedule 40 steel pipe carrying 650 gal/min of SAE 10W oil at 100°F.

8–15   Determine the pressure drop across a 90° standard elbow in a $2\frac{1}{2}$-inch Schedule 40 steel pipe if water at 60° F is flowing at the rate of 200 gal/min.

8–16   Determine the energy loss which occurs as 10 gal/min of water at 60°F flows around a 90° bend in a commercial steel tube having an outside diameter of $\frac{3}{4}$ inch and a wall thickness of 0.065 inch. The radius of the bend to the centerline of the tube is 6 inches.

8–17   Figure 8–21 shows a test setup to determine the energy loss due to a heat exchanger. Water at 120°F is flowing vertically upward at 0.20 ft³/sec. The flow area at section 1 is 3 in.² and at section 2 is 12 in.² Calculate the energy loss between points 1 and 2. Determine the loss coefficient for the heat exchanger based on the velocity in the inlet pipe.

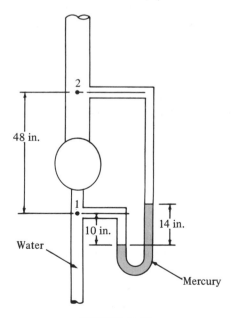

FIGURE 8–21

8–18 Figure 8–22 depicts gasoline flowing from a storage tank into a truck for transport. The gasoline has a specific gravity of 0.721 and the temperature is 77°F. Determine the required depth h in the tank to produce a flow rate of 350 gal/min into the truck. Since the pipes are short, neglect the energy losses due to pipe friction but do consider minor losses.

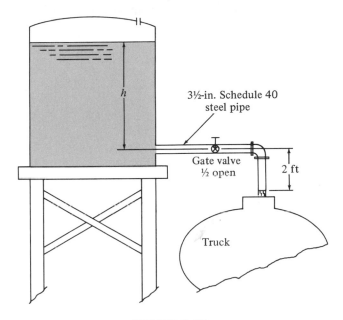

3½-in. Schedule 40 steel pipe

Gate valve ½ open

2 ft

Truck

FIGURE 8–22

9

Series Pipe Line Systems

System Classifications

Most pipe flow systems involve both major friction energy losses and
minor losses. If the system is arranged so that the fluid flows through a
continuous line without branching, it is referred to as a series system.
Conversely, if the flow branches into two or more lines it is referred to as
a parallel system.

This chapter deals only with series systems such as the one illustrated
in Fig. 9–1. If the energy equation is written for this system, using the

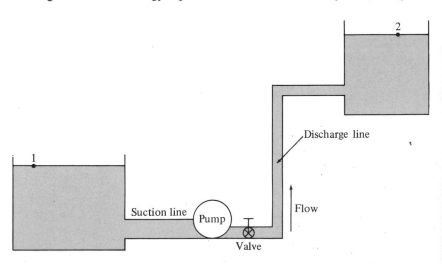

FIGURE 9–1

222

surface of each reservoir as the reference points, it would appear as shown below.

$$\frac{p_1}{\gamma} + z_1 + \frac{v_1{}^2}{2g} + h_A - h_L = \frac{p_2}{\gamma} + z_2 + \frac{v_2{}^2}{2g} \qquad \text{(9–1)}$$

The first three terms on the left side of this equation represent the energy possessed by the fluid at point 1 in the form of pressure head, elevation head, and velocity head. Likewise the three terms on the right side of the equation represent the energy possessed by the fluid at point 2. The two terms, h_A and h_L, indicate the energy added to the fluid and the energy lost from the system anywhere between the reference points 1 and 2. In this problem h_A is the energy added by the pump. Energy is lost, however, due to several conditions. We can say that

$$h_L = h_1 + h_2 + h_3 + h_4 + h_5 + h_6 \qquad \text{(9–2)}$$

where

h_L = Total energy loss per pound of fluid flowing
h_1 = Entrance loss
h_2 = Friction loss in the suction line
h_3 = Energy loss in the valve
h_4 = Energy loss in the two 90° elbows
h_5 = Friction loss in the discharge line
h_6 = Exit loss

In a series pipe line the total energy loss is the sum of the individual major and minor losses. This statement is in agreement with the principle that the use of the energy equation is a means of accounting for all of the energy in the system between the two reference points.

When designing or analyzing a pipe flow system there are six primary parameters involved, namely,

1. The energy losses from the system or energy additions to the system
2. The volume flow rate of the fluid or the velocity of flow
3. The size of the pipe
4. The length of pipe
5. The pipe wall roughness, ϵ
6. The fluid properties of specific weight, density, and viscosity

Usually, one of the first three parameters is to be determined while the remaining items are known or can be specified by the designer. The method of performing the design or completing the analysis is different depending on what is unknown. The methods described in this book are classified as follows.

Class I : The energy losses or additions are to be determined
Class II : The volume flow rate is to be determined
Class III: The pipe diameter is to be determined

9–2   Class I Systems

The approach to the analysis of Class I systems is identical to that used throughout the previous chapters except that generally many types of energy losses will exist and each energy loss must be evaluated and included in the energy equation. The total energy loss, h_L, is the sum of the individual major and minor losses. The following programmed example problem will illustrate the solution of a Class I problem.

Programmed Example Problem

Example Problem 9–1.   Calculate the power supplied to the pump shown in Fig. 9–2 if its efficiency is 76 percent. Methyl alcohol at 77°F is flowing at the rate of 0.50 ft³/sec. The suction line is a standard 4-inch Schedule 40 steel pipe, 50 feet long. The total length of 2-inch Schedule 40 steel pipe in the discharge line is 650 feet. Assume that the entrance from reservoir 1 is through a square-edged inlet and that the elbows are standard. The valve is a fully open globe valve.

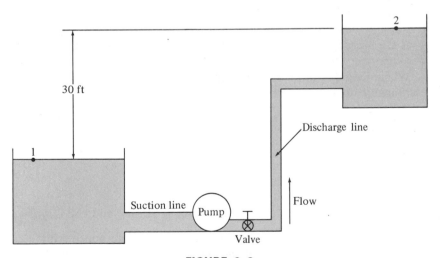

FIGURE 9–2

To begin the solution, write the energy equation for the system.

Using the surfaces of the reservoirs as the reference points you should have

$$\frac{p_1}{\gamma} + z_1 + \frac{v_1^2}{2g} + h_A - h_L = \frac{p_2}{\gamma} + z_2 + \frac{v_2^2}{2g}$$

Since $p_1 = p_2 = 0$ and v_1 and v_2 are approximately zero the equation can be simplified to

$$z_1 + h_A - h_L = z_2$$

Since the objective of the problem is to calculate the power supplied to the pump, solve now for the total head on the pump.

The total head is

$$h_A = z_2 - z_1 + h_L$$

There are six components to the total energy loss. List them and write the formula for evaluating each one.

Your list should include the following items where the subscript s indicates the suction line and the subscript d indicates the discharge line.

$$h_1 = C_L \frac{v_s^2}{2g} \qquad\qquad \text{(Entrance loss)}$$

$$h_2 = f_s \frac{L}{D} \frac{v_s^2}{2g} \qquad\qquad \text{(Friction loss in suction line)}$$

$$h_3 = f_d \frac{L_e}{D} \frac{v_d^2}{2g} \qquad\qquad \text{(Valve)}$$

$$h_4 = 2f_d \frac{L_e}{D} \frac{v_d^2}{2g} \qquad\qquad \text{(Two 90° elbows)}$$

$$h_5 = f_d \frac{L}{D} \frac{v_d^2}{2g} \qquad\qquad \text{(Friction loss in discharge line)}$$

$$h_6 = 1.0 \frac{v_d^2}{2g} \qquad\qquad \text{(Exit loss)}$$

Since the velocity head in the suction or discharge line is required for each energy loss it is convenient to calculate these values now.

You should have $v_s^2/2g = 0.498$ ft and $v_d^2/2g = 7.15$ ft.

$$v_s = \frac{Q}{A_s} = \frac{0.50 \text{ ft}^3}{\text{sec}} \cdot \frac{1}{0.0884 \text{ ft}^2} = 5.66 \text{ ft/sec}$$

$$\frac{v_s^2}{2g} = \frac{(5.66)^2}{2(32.2)} \text{ ft} = 0.498 \text{ ft}$$

$$v_d = \frac{Q}{A_d} = \frac{0.50 \text{ ft}^3}{\text{sec}} \cdot \frac{1}{0.02333 \text{ ft}^2} = 21.4 \text{ ft/sec}$$

$$\frac{v_d^2}{2g} = \frac{(21.4)^2}{2(32.2)} \text{ ft} = 7.15 \text{ ft}$$

The values for the Reynolds number in each pipe are also required to determine the friction losses. Calculate these now.

For the suction line $N_R = 2.48 \times 10^5$ and $N_R = 4.83 \times 10^5$ for the discharge line.

$$N_R = vD\rho/\mu$$

For methyl alcohol at 77°F, $\rho = 1.53$ slugs/ft³ and $\mu = 1.17 \times 10^{-5}$ lb-sec/ft². Then in the suction line,

$$N_R = \frac{vD\rho}{\mu} = \frac{(5.66)(0.3355)(1.53)}{1.17 \times 10^{-5}} = 2.48 \times 10^5$$

In the discharge line,

$$N_R = \frac{vD\rho}{\mu} = \frac{(21.4)(0.1723)(1.53)}{1.17 \times 10^{-5}} = 4.83 \times 10^5$$

Therefore the flow in both lines is turbulent.

Returning now to the energy loss calculations, evaluate h_1, the entrance loss, in ft-lb/lb.

The result is $h_1 = 0.249$ ft. For a square-edged inlet, $C_L = 0.5$ and

$$h_1 = 0.5 \, (v_s^2/2g) = (0.5)(0.498) \text{ ft} = 0.249 \text{ ft}$$

Now calculate h_2.

$h_2 = 1.34$ ft is the result.

$$h_2 = f_s \frac{L}{D} \frac{v_s^2}{2g} = f_s \frac{50}{0.3355}(0.498) \text{ ft}$$

The value of f_s must be evaluated from the Moody diagram, Fig. 7–2. For steel pipe, $\epsilon = 1.5 \times 10^{-4}$ ft.

$$D/\epsilon = 0.3355/1.5 \times 10^{-4} = 0.224 \times 10^4 = 2240$$
$$N_R = 2.48 \times 10^5$$

Then $f_s = 0.018$ and

$$h_2 = \frac{(0.018)(50)(0.498)}{0.3355} \text{ ft} = 1.34 \text{ ft}$$

Now calculate h_3.

From the data in Chapter 8 the equivalent length ratio, L_e/D, for a fully open globe valve is 340. Evaluating the friction factor f_d gives

$$D/\epsilon = 0.1723/1.5 \times 10^{-4} = 0.115 \times 10^4 = 1150$$
$$N_R = 4.83 \times 10^5$$
$$f_d = 0.0185$$

Then,

$$h_3 = f_d \frac{L_e}{D} \frac{v_d^2}{2g} = (0.0185)(340)(7.15) \text{ ft} = 45.0 \text{ ft}$$

Now calculate h_4.

For standard 90° elbows, $L_e/D = 30$. The value of f_d is the same as that found in the preceding panel. Then,

$$h_4 = 2f_d \frac{L_e}{D} \frac{v_d^2}{2g} = (2)(0.0185)(30)(7.15) \text{ ft} = 7.93 \text{ ft}$$

Now calculate h_5.

For the discharge line friction loss,

$$h_5 = f_d \frac{L}{D} \frac{v_d^2}{2g} = \frac{(0.0185)(650)(7.15)}{0.1723} \text{ ft} = 499 \text{ ft}$$

Now calculate h_6.

The exit loss is

$$h_6 = 1.0 \, (v_d^2/2g) = 7.15 \text{ ft}$$

This concludes the calculation of the individual energy losses. The total loss, h_L, can now be determined.

$$h_L = h_1 + h_2 + h_3 + h_4 + h_5 + h_6$$
$$h_L = (0.249 + 1.34 + 45.0 + 7.93 + 499 + 7.15) \text{ ft}$$
$$h_L = 561 \text{ ft}$$

From the energy equation the expression for the total head on the pump was found to be

$$h_A = z_2 - z_1 + h_L$$

Then

$$h_A = 30 \text{ ft} + 561 \text{ ft} = 591 \text{ ft}$$

Now calculate the power supplied to the pump.

$$\text{Power} = \frac{h_A \gamma Q}{e_M} = \frac{(591)(49.10)(0.50)}{(550)(0.76)} \text{ horsepower}$$

$$\text{Power} = 34.8 \text{ horsepower}$$

This concludes the programmed example problem.

===

9–3   **Class II Systems**

Whenever the volume flow rate in the system is unknown the analysis of the system performance is done by a procedure called iteration. This is required because there are too many unknown quantities to use the direct solution procedure described for Class I problems. Specifically, if the volume flow rate is unknown then the velocity of flow is also unknown. It follows that the Reynolds number is unknown since it depends on the velocity. If the Reynolds number cannot be found then the friction factor f cannot be determined directly. Since energy losses due to friction are dependent on both the velocity and the friction factor, the value of these losses cannot be calculated directly.

Iteration overcomes these difficulties. It is a type of trial and error solution method in which a trial value is assumed for the unknown friction factor, f, allowing the calculation of a corresponding velocity of flow. The procedure provides a means of checking the accuracy of the trial value of f and also indicates the new trial value to be used if an additional calculation cycle is required. This procedure for solving Class II problems is presented in step-by-step form below. The two programmed example problems which follow then illustrate the application of the procedure.

Class II Systems with One Pipe — Solution Procedure

1. Write the energy equation for the system.
2. Evaluate known quantities such as pressure heads and elevation heads.
3. Express energy losses in terms of the unknown velocity, v, and friction factor, f.
4. Solve for the velocity in terms of f.
5. Express the Reynolds number in terms of the velocity.
6. Calculate the relative roughness D/ϵ.
7. Select a trial value of f based on the known D/ϵ and a Reynolds number in the turbulent range.
8. Calculate the velocity using the equation from Step 4.
9. Calculate the Reynolds number from the equation in Step 5.
10. Evaluate the friction factor f for the Reynolds number from Step 9 and the known value of D/ϵ, using the Moody diagram, Fig. 7–2.
11. If the new value of f is different from the value used in Step 8, repeat Steps 8 through 11 using the new value of f.
12. If there is no significant change in f from the assumed value, then the velocity found in Step 8 is correct.

Programmed Example Problem

Example Problem 9–2. It is desired to pump a lubricating oil through a horizontal six-inch Schedule 40 steel pipe with a maximum pressure drop of 2.5 psi per 100 feet of pipe. The oil has a specific gravity of 0.88 and a dynamic viscosity of 2.0×10^{-4} lb-sec/ft^2. Calculate the maximum allowable volume flow rate of oil.

Is this a Class II system?

Yes, it is. Since the volume flow rate is unknown the Class II solution procedure described above must be used. To begin, draw a sketch of the system, write the energy equation, and simplify it as much as possible.

Figure 9–3 shows the two points of interest in the pipe. Then the energy equation is

$$\frac{p_1}{\gamma} + z_1 + \frac{v_1{}^2}{2g} - h_L = \frac{p_2}{\gamma} + z_2 + \frac{v_2{}^2}{2g}$$

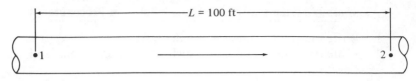

$$p_1 - p_2 \leq 2.5 \text{ psi}$$

FIGURE 9–3

But $z_1 = z_2$ and $v_1 = v_2$. Then,

$$\frac{p_1}{\gamma} - h_L = \frac{p_2}{\gamma}$$

$$\frac{p_1 - p_2}{\gamma} = h_L$$

Evaluate the pressure head difference now.

You should have $(p_1 - p_2)/\gamma = 6.56$ ft. Since the allowable pressure drop is 2.5 psi,

$$\frac{p_1 - p_2}{\gamma} = \frac{2.5 \text{ lb}}{\text{in.}^2} \cdot \frac{\text{ft}^3}{(0.88)(62.4) \text{ lb}} \cdot \frac{144 \text{ in.}^2}{\text{ft}^2} = 6.56 \text{ ft}$$

Then $h_L = 6.56$ ft. The total energy loss in this case is due to friction. From Darcy's equation,

$$h_L = f \frac{L}{D} \frac{v^2}{2g}$$

Step 4 of the solution procedure says to express the velocity in terms of the friction factor. Do this now.

The final form should be $v = \sqrt{2.14/f}$.

$$h_L = f \frac{L}{D} \frac{v^2}{2g}$$

$$v = \sqrt{\frac{2gh_L D}{Lf}}$$

In this equation, $g = 32.2 \text{ ft/sec}^2$, $h_L = 6.56$ ft, $D = 0.5054$ ft, and $L = 100$ ft. Then,

$$v = \sqrt{\frac{(2)(32.2)(6.56)(0.5054)}{(100)(f)}} = \sqrt{\frac{2.14}{f}}$$

Now do Step 5 of the procedure.

You should have $N_R = (0.433 \times 10^4)v$.

$$N_R = \frac{vD\rho}{\mu}$$

But $D = 0.5054$ ft, $\mu = 2.0 \times 10^{-4}$ lb-sec/ft², and

$$\rho = (0.88)(1.94 \text{ slugs/ft}^3) = 1.71 \text{ slugs/ft}^3$$

Then,

$$N_R = \frac{v(0.5054)(1.71)}{2.0 \times 10^{-4}} = (0.433 \times 10^4)v$$

Now do Step 6.

Since $\epsilon = 1.5 \times 10^{-4}$ for steel pipe,

$$D/\epsilon = (0.5054)/(1.5 \times 10^{-4}) = 3380$$

These first six steps are all preliminary to the iteration portion of the procedure, Steps 7–12. The results found above will simplify the process of iteration.

Step 7 states that a trial value of the friction factor f should be selected. Moody's diagram, Fig. 7–2, can be used as an aid to the rational selection of a trial value. Since the relative roughness is known to be 3380, the range of possible values of f is from approximately 0.039 for $N_R = 4000$, to 0.015 for $N_R = 1.0 \times 10^7$ and larger. Any value in this range can be selected for the first trial. Use $f = 0.020$ and proceed with Step 8.

For $f = 0.020$, the velocity would be

$$v = \sqrt{2.14/f} = \sqrt{2.14/0.020} = \sqrt{107}$$
$$v = 10.35 \text{ ft/sec}$$

We can now calculate the corresponding Reynolds number.

$$N_R = (0.433 \times 10^4)v = (0.433 \times 10^4)(10.35)$$
$$N_R = 4.48 \times 10^4$$

For this value of the Reynolds number and $D/\epsilon = 3380$, the new value of $f = 0.0225$. Since this is different from the initially assumed value, Steps 8 through 11 must be repeated. Do Steps 8, 9, and 10 before looking at the next panel.

The results are

$$v = \sqrt{\frac{2.14}{0.0225}} = \sqrt{95.1} = 9.76 \text{ ft/sec}$$

$$N_R = (0.433 \times 10^4)(9.76) = 4.23 \times 10^4$$

The new value of f is 0.0225, unchanged from the previous value. Therefore, $v = 9.76$ ft/sec is the correct velocity. Now the volume flow rate can be calculated to complete the problem.

$$Q = Av = (0.2006 \text{ ft}^2)(9.76 \text{ ft/sec}) = 1.96 \text{ ft}^3/\text{sec}$$

This programmed example problem is concluded.

The problem presented in the next panel is another example of a Class II system having one pipe. However, there are minor losses in addition to the pipe friction loss. The details of the solution are slightly different but the general solution procedure is the same as that used for Example Problem 9–2.

Example Problem 9–3. Water at 80°F is being supplied to an irrigation ditch from an elevated storage reservoir as shown in Fig. 9–4. Calculate the volume flow rate of water into the ditch.

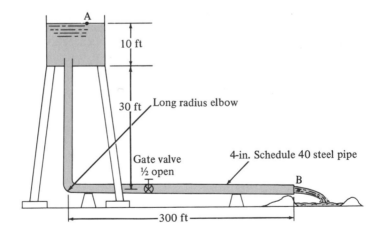

Begin with Step 1 of the solution procedure by writing the energy equation. Use points A and B as the reference points and simplify the equation as much as possible.

Compare this with your solution.

$$\frac{p_A}{\gamma} + z_A + \frac{v_A^2}{2g} - h_L = \frac{p_B}{\gamma} + z_B + \frac{v_B^2}{2g}$$

Since $p_A = p_B = 0$, and v_A is approximately zero,

$$z_A - h_L = z_B + (v_B^2/2g)$$
$$z_A - z_B = (v_B^2/2g) + h_L \qquad \text{(9–3)}$$

Notice that the stream of water at point B has the same velocity as that inside the pipe.

The elevation difference, $z_A - z_B$, is known to be 40 feet. However, the energy losses which make up h_L all depend on the unknown velocity, v_B. Thus, iteration is required. Do Step 3 of the solution procedure now.

There are four components of the total energy loss, h_L.

$$h_L = h_1 + h_2 + h_3 + h_4$$

where

$$
\begin{array}{ll}
h_1 = 1.0\,(v_B^2/2g) & \text{(Entrance loss)} \\
h_2 = f(L/D)(v_B^2/2g) & \text{(Pipe friction loss)} \\
\quad = f(330/0.3355)(v_B^2/2g) & \\
\quad = 985f(v_B^2/2g) & \\
h_3 = f(L_e/D)(v_B^2/2g) & \text{(Long radius elbow)} \\
\quad = 20f(v_B^2/2g) & \\
h_4 = f(L_e/D)(v_B^2/2g) & \text{(Half-open gate valve)} \\
\quad = 160f(v_B^2/2g) &
\end{array}
$$

Then,

$$h_L = (1.0 + 985f + 20f + 160f)(v_B^2/2g)$$
$$h_L = (1.0 + 1165f)(v_B^2/2g) \qquad \text{(9–4)}$$

Now substitute this expression for h_L into Eq. (9–3) and solve for v_B in terms of f.

You should have $v_B = \sqrt{2580/(2.0 + 1165\,f)}$.

$$z_A - z_B = (v_B^2/2g) + h_L$$
$$40 \text{ ft} = (v_B^2/2g) + (1.0 + 1165f)(v_B^2/2g)$$
$$40 \text{ ft} = (2.0 + 1165f)(v_B^2/2g)$$

Solving for v_B,

$$v_B = \sqrt{\frac{2g(40)}{2.0 + 1165f}} = \sqrt{\frac{2580}{2.0 + 1165f}} \qquad (9\text{–}5)$$

Equation (9–5) represents the completion of Step 4 of the procedure. Now do Steps 5 and 6.

$$N_R = \frac{v_B D}{\nu} = \frac{v_B(0.3355)}{9.15 \times 10^{-6}} = (0.366 \times 10^5)v_B \qquad (9\text{–}6)$$

$$D/\epsilon = (0.3355/1.5 \times 10^{-4}) = 2235$$

Step 7 is the start of the iteration process. What is the possible range of values for the friction factor for this system?

Since $D/\epsilon = 2235$, the lowest possible value of f is 0.0155 for very high Reynolds numbers and the highest possible value is 0.039 for a Reynolds number of 4000. The initial trial value of f must be in this range. Use $f = 0.020$ and complete Steps 8 and 9.

The values for velocity and Reynolds number are found using Eqs. (9–5) and (9–6).

$$v_B = \sqrt{\frac{2580}{2.0 + (1165)(0.02)}} = \sqrt{102} = 10.1 \text{ ft/sec}$$

$$N_R = (0.366 \times 10^5)(10.1) = 3.70 \times 10^5$$

Now do Step 10.

You should have $f = 0.0175$. Since this is different from the initial trial value of f, Steps 8 through 11 must be repeated now.

Using $f = 0.0175$,

$$v_B = \sqrt{\frac{2580}{2.0 + (1165)(0.0175)}} = \sqrt{115} = 10.7 \text{ ft/sec}$$

$$N_R = (0.366 \times 10^5)(10.7) = 3.92 \times 10^5$$

The new value of f is 0.0175, unchanged. Therefore,

$$v_B = 10.7 \text{ ft/sec}$$
$$Q = A_B v_B = (0.0884 \text{ ft}^2)(10.7 \text{ ft/sec}) = 0.946 \text{ ft}^3/\text{sec}$$

This programmed example problem is concluded.

Another Class II system will now be presented which is more complex then the two just completed. It includes minor losses in addition to friction losses, and has two pipes of different sizes in series. These factors require that the solution procedure be modified. Since there are two pipes, there are two unknown friction factors and two unknown velocities. Although more calculations are required, the solution procedure presented below is a straightforward iteration process similar to that used before. Under average conditions for pipe flow, the procedure will yield the final result in two iteration cycles.

Class II Systems with Two Pipes — Solution Procedure

1. Write the energy equation for the system.
2. Evaluate known quantities such as pressure heads and elevation heads.
3. Express energy losses in terms of the two unknown velocities and the two friction factors.
4. Using the continuity equation, express the velocity in the smaller pipe in terms of that in the larger pipe.

$$A_1 v_1 = A_2 v_2$$
$$v_1 = v_2(A_2/A_1)$$

5. Substitute the expression in Step 4 into the energy equation, thereby eliminating one unknown velocity.
6. Solve for the remaining velocity in terms of the two friction factors.
7. Express the Reynolds number for each pipe in terms of the velocity in that pipe.
8. Calculate the relative roughness D/ϵ for each pipe.
9. Select trial values for f in each pipe using the known values of D/ϵ as a guide. In general, the two friction factors will not be equal.
10. Calculate the velocity in the larger pipe using the equation from Step 6.
11. Calculate the velocity in the smaller pipe using the equation from Step 4.

12. Calculate the two Reynolds numbers.
13. Determine the new value of the friction factor for each pipe.
14. Compare the new values of f with those assumed in Step 9 and repeat Steps 9–14 until no significant changes can be detected. The velocities found in Steps 10 and 11 are then correct.

The programmed example problem below will illustrate the details of applying this procedure.

Programmed Example Problem

Example Problem 9–4. The piping system shown in Fig. 9–5 is being used to transfer water at 60°F from one storage tank to another. Determine the volume flow rate of water through the system. The larger pipe is a standard 6-inch Schedule 40 steel pipe having a total length of 100 feet. The smaller pipe is a standard 2-inch Schedule 40 steel pipe having a total length of 50 feet. The elbows are standard.

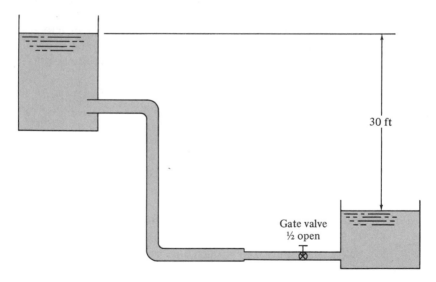

30 ft

Gate valve
½ open

FIGURE 9–5

Perform the first three steps of the solution procedure before looking at the next panel.

The correct results are summarized below. The subscripts A and B refer to the surfaces of the upper and lower tanks, respectively. The subscripts 6 and 2 refer to the 6-inch and 2-inch pipes, respectively.

Energy equation:

$$\frac{p_A}{\gamma} + z_A + \frac{v_A^2}{2g} - h_L = \frac{p_B}{\gamma} + z_B + \frac{v_B^2}{2g}$$

But $p_A = p_B = 0$ and v_A and v_B are approximately zero.

$$z_A - z_B = h_L = 30 \text{ ft} \tag{9-7}$$

Energy losses:

$$h_L = h_1 + h_2 + h_3 + h_4 + h_5 + h_6 + h_7$$
$$h_1 = 1.0 \ (v_6^2/2g) \qquad\qquad\qquad\qquad \text{(Entrance loss)}$$
$$h_2 = f_6(L/D)(v_6^2/2g) \qquad\qquad \text{(Friction in 6-inch pipe)}$$
$$\quad\ = f_6(100/0.5054)(v_6^2/2g) = 198f_6(v_6^2/2g)$$
$$h_3 = 2f_6(L_e/D)(v_6^2/2g) \qquad\qquad \text{(Two 90° elbows)}$$
$$\quad\ = 60 \ f_6(v_6^2/2g)$$
$$h_4 = C_L(v_2^2/2g) \qquad\qquad\qquad \text{(Sudden contraction)}$$

Using $D_6/D_2 = 0.5054/0.1723 = 2.93$ and assuming that v_2 is approximately 10 ft/sec, we find $C_L = 0.42$ from the data in Chapter 8. If v_2 is much different from 10 ft/sec, this value of C_L may have to be reevaluated.

$$h_4 = 0.42(v_2^2/2g)$$
$$h_5 = f_2(L/D)(v_2^2/2g) \qquad\qquad \text{(Friction in 2-inch pipe)}$$
$$\quad\ = f_2(50/0.1723)(v_2^2/2g) = 290 \ f_2(v_2^2/2g)$$
$$h_6 = f_2(L_e/D)(v_2^2/2g) \qquad\qquad \text{(Gate valve — } \tfrac{1}{2} \text{ open)}$$
$$\quad\ = 160 \ f_2(v_2^2/2g)$$
$$h_7 = 1.0 \ (v_2^2/2g) \qquad\qquad\qquad\qquad \text{(Exit loss)}$$

It is convenient to algebraically sum the energy losses as shown below.

$$h_L = (1.0 + 198f_6 + 60f_6)(v_6^2/2g)$$
$$\qquad\qquad\qquad\qquad + (0.42 + 290f_2 + 160f_2 + 1.0)(v_2^2/2g)$$
$$h_L = (1.0 + 258f_6)(v_6^2/2g) + (1.42 + 450f_2)(v_2^2/2g) \tag{9-8}$$

This is the simplest form of the expression for the energy loss. Now do Step 4 of the solution procedure.

You should have the following:

$$v_2 = v_6(A_6/A_2) = v_6(0.2006/0.02333)$$
$$v_2 = 8.61 \ v_6 \tag{9-9}$$

In order to substitute this into Eq. (9–8), the expression should be squared.

$$v_2^2 = 74.2\, v_6^2 \qquad\qquad\qquad \textbf{(9–10)}$$

Now do Steps 5 and 6.

The final expression for v_6 is

$$v_6 = \sqrt{\frac{1930}{106 + 258f_6 + 33{,}400f_2}}$$

Here is how it was found. Substituting Eq. (9–10) into Eq. (9–8) gives

$$h_L = (1.0 + 258f_6)(v_6^2/2g) + (1.42 + 450f_2)(74.2\, v_6^2/2g)$$
$$h_L = (1.0 + 258f_6 + 105 + 33{,}400f_2)(v_6^2/2g)$$
$$h_L = (106 + 258f_6 + 33{,}400f_2)(v_6^2/2g)$$

Solving for v_6,

$$v_6 = \sqrt{\frac{2gh_L}{106 + 258f_6 + 33{,}400f_2}}$$

From Eq. (9–7), $h_L = 30$ ft. Then,

$$v_6 = \sqrt{\frac{1930}{106 + 258f_6 + 33{,}400f_2}} \qquad\qquad \textbf{(9–11)}$$

Now do Steps 7 and 8 of the solution procedure.

Using $\nu = 1.21 \times 10^{-5}$ ft^2/sec for water at 60° and $\epsilon = 1.5 \times 10^{-4}$ ft for steel pipe,

$$(N_R)_6 = (0.417 \times 10^5)v_6$$
$$(N_R)_2 = (0.142 \times 10^5)v_2$$
$$(D/\epsilon)_6 = 3370$$
$$(D/\epsilon)_2 = 1150$$

Step 9 begins the iteration procedure. As initial trial values for the friction factors, use $f_6 = 0.02$ and $f_2 = 0.025$. These are in the range of possible values for f for the known relative roughness values. Now do Steps 10 through 13 as the first cycle of iteration.

Here are the correct solutions.

$$v_6 = \sqrt{\frac{1930}{106 + (258)(0.02) + 33,400(0.025)}} = \sqrt{2.04} = 1.43 \text{ ft/sec}$$

$$v_2 = 8.61\ v_6 = 12.3 \text{ ft/sec}$$

$$(N_R)_6 = (0.417 \times 10^5)(1.43) = 5.96 \times 10^4$$

$$(N_R)_2 = (0.142 \times 10^5)(12.3) = 1.75 \times 10^5$$

Then the new $f_6 = 0.0215$ and the new $f_2 = 0.020$. Since these are different from the initially assumed values, repeat Steps 10–13.

These are the results.

$$v_6 = 1.57 \text{ ft/sec}$$

$$v_2 = 13.5 \text{ ft/sec}$$

$$(N_R)_6 = 6.55 \times 10^4$$

$$(N_R)_2 = 1.92 \times 10^5$$

The values of f_6 and f_2 are unchanged. Therefore, the velocities listed above are correct. Using v_6 to calculate the volume flow rate gives

$$Q = A_6 v_6 = (0.2006)(1.57) \text{ ft}^3/\text{sec} = 0.314 \text{ ft}^3/\text{sec}$$

9–4   **Class III Systems**

Systems which fall into Class III present true design problems. The requirements placed on the system are specified in terms of an allowable pressure drop or energy loss, a desired volume flow rate, the fluid properties, and the type of pipe which is to be used. Then the proper size pipe is determined which will meet these requirements.

Iteration is required to solve Class III system design problems since there are too many unknowns to allow a direct solution. The flow velocity, Reynolds number, and the relative roughness, D/ϵ, are all dependent on the pipe diameter. Therefore, the friction factor cannot be determined directly.

The procedure for designing Class III systems is different depending on the complexity of the system. The simplest case is when only pipe friction loss is to be considered. A step-by-step solution procedure is given below for this kind of system along with an example problem. Systems which include minor losses, in which loss coefficients, C_L, or equivalent length ratios, L_e/D, are given, are more complex. A solution procedure for this type of system is demonstrated through another example problem.

Class III Systems with Pipe Friction Loss Only

The problem is to select the proper size pipe which will carry a given volume flow rate of a fluid with a certain maximum allowable pressure drop. The solution procedure is outlined below. The first seven steps represent an algebraic reduction of the problem to a simple form. Steps 8 through 13 comprise the iteration routine.

1. Write the energy equation for the system.
2. Solve for the total energy loss, h_L, and evaluate known pressure heads and elevations.
3. Express the energy loss in terms of the velocity using Darcy's equation.

$$h_L = f \frac{L}{D} \frac{v^2}{2g}$$

4. Express the velocity in terms of the volume flow rate and the pipe diameter.

$$v = \frac{Q}{A} = \frac{4Q}{\pi D^2}$$

5. Substitute the expression for v into Darcy's equation.

$$h_L = f \frac{L}{D} \frac{16Q^2}{\pi^2 D^4 (2g)} = \frac{8LQ^2}{\pi^2 g} \frac{f}{D^5}$$

6. Solve for the diameter.

$$D = \left[\frac{8LQ^2}{\pi^2 g h_L} f \right]^{1/5} = (C_1 f)^{0.2}$$

Notice that the terms which make up C_1 are all known and are independent of the pipe diameter.

7. Express the Reynolds number in terms of the diameter.

$$N_R = \frac{vD\rho}{\mu} = \frac{vD}{\nu}$$

But $v = 4Q/\pi D^2$. Then,

$$N_R = \frac{4Q}{\pi D^2} \frac{D}{\nu} = \frac{4Q}{\pi \nu} \frac{1}{D} = \frac{C_2}{D}$$

where $C_2 = 4Q/\pi \nu$.

8. Assume an initial trial value for f. Since both N_R and D/ϵ are unknown, there are no specific guidelines for selecting the initial value. Unless special conditions exist, or experience dictates otherwise, assume $f = 0.02$.
9. Compute $D = (C_1 f)^{0.2}$.

10. Compute $N_R = (C_2/D)$.
11. Compute D/ϵ.
12. Determine a new value for the friction factor, f, from Moody's diagram, Fig. 7-2.
13. Compare the new value of f with that assumed in Step 8 and repeat Steps 8–12 until no significant change in f can be detected. The diameter calculated in Step 9 is then correct.

The programmed example problem which follows illustrates the application of this procedure.

Programmed Example Problem

Example Problem 9–5. A water line is to be run to the green of the seventh hole of a golf course as shown in Fig. 9–6. The supply is from a main line at point A where the pressure is 80 psig. In order to ensure proper operation of the sprinklers at the green, the pressure at point B must be at least 60 psig. Determine the smallest allowable size of standard Schedule 40 steel pipe to supply 0.50 ft³/sec of water at 60°F.

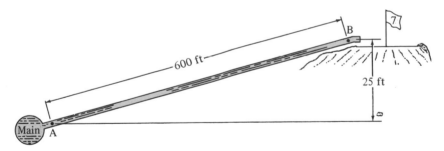

FIGURE 9–6

Do Steps 1 and 2 of the Class III system solution procedure now to determine the allowable energy loss, h_L.

You should have $h_L = 21.2$ ft.

Energy equation:

$$\frac{p_A}{\gamma} + z_A + \frac{v_A^2}{2g} - h_L = \frac{p_B}{\gamma} + z_B + \frac{v_B^2}{2g}$$

But $v_A = v_B$. Then,

$$h_L = \frac{p_A - p_B}{\gamma} + z_A - z_B$$

$$= \frac{(80 - 60)\ \text{lb}}{\text{in.}^2} \cdot \frac{\text{ft}^3}{62.4\ \text{lb}} \cdot \frac{144\ \text{in.}^2}{\text{ft}^2} - 25\ \text{ft}$$

$$h_L = (46.2 - 25)\ \text{ft} = 21.2\ \text{ft}$$

It is now required to determine the proper size pipe which will limit the energy loss due to friction to no more than 21.2 feet. Steps 3, 4, and 5 represent the reduction of Darcy's equation. Since we are only concerned with friction losses in this problem, these steps are identical. Therefore, go to Step 6 and evaluate the constant, C_1.

You should get $C_1 = 0.178$ and $D = (0.178\ f)^{0.2}$.

$$C_1 = \frac{8LQ^2}{\pi^2 gh_L} = \frac{(8)(600)(0.50)^2}{\pi^2\ (32.2)(21.2)} = 0.178$$

Now evaluate C_2 in Step 7.

The correct value is $C_2 = 0.526 \times 10^5$.

$$C_2 = \frac{4Q}{\pi \nu} = \frac{(4)(0.5)}{\pi(1.21 \times 10^{-5})} = 0.526 \times 10^5$$

Then $N_R = (0.526 \times 10^5)/D$. Now, using $f = 0.02$ for the initial trial, complete Steps 8–12 before looking at the next panel.

Compare your results with these.

$$f = 0.02 \qquad\qquad\qquad\qquad\qquad\qquad\qquad\qquad \text{(assumed)}$$
$$D = [(0.178)(0.02)]^{0.2} = (0.00356)^{0.2} = 0.324\ \text{ft}$$
$$N_R = (0.526 \times 10^5)/(0.324) = 1.625 \times 10^5$$
$$D/\epsilon = (0.324)/(1.5 \times 10^{-4}) = 2160 \qquad\qquad \text{(Steel pipe)}$$
$$\text{New}\ f = 0.019$$

Since the new value for f is different from the assumed value, repeat Steps 8–12 now.

Here are the revised values using $f = 0.019$.

$$D = 0.320\ \text{ft}$$
$$N_R = 1.65 \times 10^5$$
$$D/\epsilon = 2135$$

The new value of $f = 0.019$ is unchanged. Therefore, the minimum allowable size for the pipe is 0.320 feet. Now select a standard Schedule 40 pipe closest to this size.

A nominal 4-inch, Schedule 40 steel pipe has an internal diameter of 0.3355 feet and is selected for this application.

This programmed example problem is concluded.

Another example problem follows in which a Class III system is to be designed. Minor losses of various kinds are included. The solution is not in programmed format but the details and the logic of the procedure should be carefully followed.

Example Problem 9–6. In a chemical processing system, propyl alcohol at 77°F is taken from the bottom of a large tank and transferred by gravity to another part of the system as shown in Fig. 9–7. The length of the line between the two tanks is 22 feet. A filter is installed in the line and it is known to have a loss coefficient, C_L, of 8.5. Drawn stainless steel tubing is to be used for the transfer line. Use $\epsilon = 1.0 \times 10^{-4}$ ft for the equivalent roughness of the tubing. Specify the standard size of tubing which would allow a volume flow rate of 40 gallons per minute through this system.

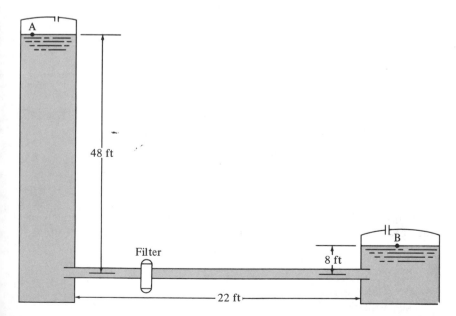

FIGURE 9–7

Solution.   Using the surfaces of the two tanks as the reference points, the energy equation is

$$\frac{p_A}{\gamma} + z_A + \frac{v_A{}^2}{2g} - h_L = \frac{p_B}{\gamma} + z_B + \frac{v_B{}^2}{2g}$$

But $p_A = p_B = 0$ and $v_A = v_B = 0$. Then,

$$h_L = z_A - z_B = 40 \text{ ft}$$

There are four components to the total energy loss.

$$h_L = 1.0 \frac{v^2}{2g} + 8.5 \frac{v^2}{2g} + f \frac{22}{D} \frac{v^2}{2g} + 1.0 \frac{v^2}{2g}$$

$$\text{(Entrance) (Filter) (Friction) (Exit)}$$

$$h_L = 10.5 \frac{v^2}{2g} + f \frac{22}{D} \frac{v^2}{2g}$$

In this equation, v is the velocity of flow in the tube. Solving for the friction factor, f, gives

$$f \frac{22}{D} \frac{v^2}{2g} = h_L - 10.5 \frac{v^2}{2g}$$

$$f = \frac{2gh_L D}{22v^2} - \frac{10.5\, v^2}{2g} \frac{2g}{22} \frac{D}{v^2}$$

$$f = \frac{2gh_L D}{22v^2} - 0.477\, D$$

Now let $v = 4Q/\pi D^2$.

$$f = \frac{2gh_L D\pi^2 D^4}{(22)(16)Q^2} - 0.477\, D$$

Since $h_L = 40$ ft and

$$Q = 40 \text{ gal/min.} \times \frac{1 \text{ ft}^3/\text{sec}}{449 \text{ gal/min}} = 0.0893 \text{ ft}^3/\text{sec}$$

$$f = \frac{(2)(32.2)(40)\,(\pi)^2}{(22)(16)(0.0893)^2} D^5 - 0.477\, D$$

$$f = 9080\, D^5 - 0.477\, D$$

This equation will be used to iterate for the diameter, D. Since we cannot solve for D in terms of f, the iteration must proceed as follows:

1. Assume a value for D.
2. Calculate f.
3. Calculate D/ϵ.
4. Calculate N_R.
5. Evaluate f and compare with that of Step 2.

6. Adjust D in a manner which will decrease the difference between the values of f and repeat Steps 2–6 until agreement is found for successive values of f.

There is a greater probability of requiring several steps of iteration with this procedure than before since judgment is required in the selection of trial values of D. Select a nominal 2-inch tube with an inside diameter of 0.15 ft as our initial trial value. Then,

$$f = (9080)(0.15)^5 - (0.477)(0.15) = 0.618$$
$$D/\epsilon = 0.15/1.0 \times 10^{-4} = 0.15 \times 10^4 = 1500$$
$$N_R = \frac{vD\rho}{\mu} = \frac{4Q\rho D}{\pi D^2 \mu} = \frac{4Q\rho}{\pi \mu}\frac{1}{D} = \frac{(4)(0.0893)(1.56)}{(\pi)(4.01 \times 10^{-5})}\frac{1}{D}$$
$$= 4.41 \times 10^3/D = 4.41 \times 10^3/0.15 = 2.94 \times 10^4$$
$$f = 0.025$$

A smaller value of D is required to cause the two values of f to be equal. The table below shows the results of successive trials for D including the one just completed.

TABLE 9–1

D	f (Step 2)	D/ϵ	N_R	f (Step 5)	Required change in D
0.15	0.618	1500	2.94×10^4	0.025	decrease
0.10	0.0431	1000	4.41×10^4	0.024	decrease
0.09	0.0105	900	4.90×10^4	0.024	increase
0.095	0.0248	950	4.65×10^4	0.024	decrease
0.094	0.0217	940	4.70×10^4	0.024	increase
0.0945	0.0235	945	4.67×10^4	0.024	O.K.

Thus the diameter of the tube should be 0.0945 foot in order to allow 40 gallons per minute of flow. The closest standard tube, from the table in the appendix, is a nominal $1\frac{1}{4}$-inch tube with a wall thickness of 0.049 inch, which has an inside diameter of 0.0960 foot.

PRACTICE PROBLEMS

9–1 Water at 50°F flows from a large reservoir at the rate of 0.5 ft³/sec through the system shown in Fig. 9–8. Calculate the pressure at B.

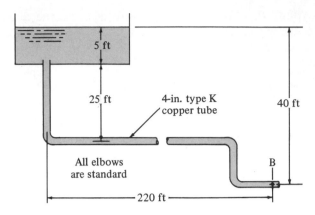

FIGURE 9–8

9–2   For the system shown in Fig. 9–9 kerosene (sg = 0.82) at 70°F
      is to be forced from tank A to reservoir B by increasing the pres-
      sure in the sealed tank A above the kerosene. The total length
      of 2-inch Schedule 40 steel pipe is 120 feet. The elbows are stand-
      ard. Calculate the required pressure in tank A to cause a flow
      rate of 115 gal/min.

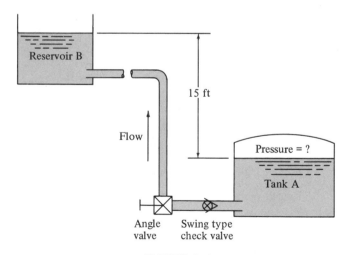

FIGURE 9–9

9–3   Figure 9–10 shows a portion of a hydraulic circuit. It is required
      that the pressure at point B is 200 psig when the volume flow
      rate is 60 gal/min. The hydraulic fluid has a specific gravity of
      0.90 and a dynamic viscosity of 6.0×10^{-5} lb-sec/ft². The total

length of pipe between A and B is 50 feet. The elbows are standard. Calculate the pressure at the outlet of the pump at A.

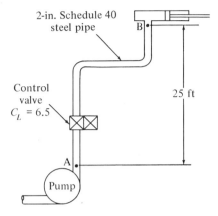

FIGURE 9–10

9–4 Figure 9–11 shows a hydraulic system in which the pressure at B must be 500 psig while the flow rate is 750 gal/min. The fluid is a medium machine tool hydraulic oil. The total length of 4-inch pipe is 40 feet. The elbows are standard. Neglect the energy loss due to friction in the 6-inch pipe. Calculate the required pressure at A if the oil is at 100°F and if it is at 210°F.

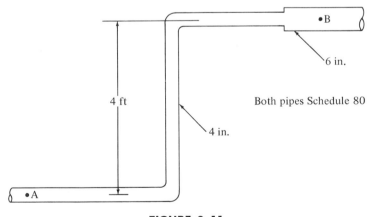

FIGURE 9–11

9–5 Oil is flowing at the rate of 0.5 ft³/sec in the system shown in Fig. 9–12. Data for the system are:

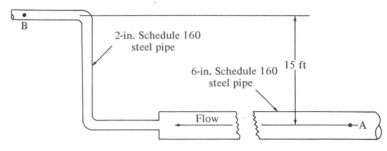

FIGURE 9–12

Oil specific weight = 56.0 lb/ft³
Oil kinematic viscosity = 2.28 × 10⁻⁴ft²/sec
Length of 6-inch pipe = 600 feet
Length of 2-inch pipe = 25 feet
Elbows are long radius type
Pressure at B is 1800 psig

Considering all pipe friction and minor losses, calculate the pressure at A.

9–6   For the system shown in Fig. 9–13 calculate the vertical distance between the surfaces of the two reservoirs when water at 50°F flows from A to B at the rate of 1.0 ft³/sec. Both pipes are cast iron and the elbows are standard. The total length of the 3-inch pipe is 300 feet and of the 6-inch pipe is 1000 feet.

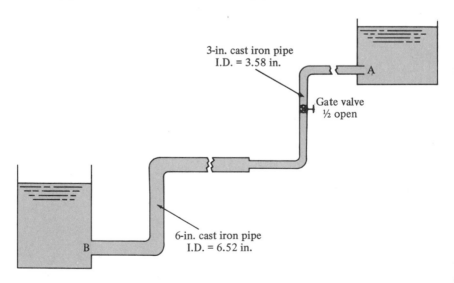

FIGURE 9–13

9–7 A liquid refrigerant flows through the system shown in Fig. 9–14 at the rate of 1.0×10^{-3} ft³/sec. The refrigerant has a specific gravity of 1.25 and a dynamic viscosity of 6×10^{-6} lb-sec/ft². Calculate the pressure difference between points A and B. The tube is steel having an outside diameter of $\frac{1}{2}$ inch, a wall thickness of 0.035 inch and a total length of 100 feet.

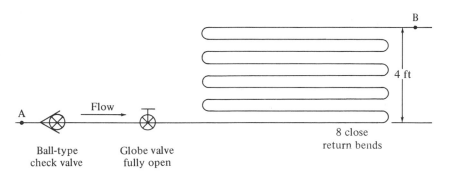

FIGURE 9–14

9–8 Water at 100°F is flowing in a 4-inch Schedule 80 steel pipe which is 25 feet long. Calculate the maximum allowable volume flow rate if the energy loss due to pipe friction is to be limited to 30 ft-lb/lb.

9–9 A hydraulic oil is flowing in a steel tube having an outside diameter of 2 inches and a wall thickness of 0.109 inch. A pressure drop of 9.8 psi is observed over 100 feet of length of the pipe. The oil has a specific gravity of 0.90 and a dynamic viscosity of 6.2×10^{-5} lb-sec/ft². Assume the tube wall roughness to be 1.0×10^{-4} feet. Calculate the velocity of flow of oil.

9–10 In a processing plant, ethylene glycol at 77°F is flowing in 5000 feet of 6-inch cast iron pipe (actual inside diameter = 6.52 inches). Over this distance the pipe falls 55 feet and the pressure drops from 250 psig to 180 psig. Calculate the velocity of flow in the pipe.

9–11 Water at 60°F is flowing downward in a vertical tube 25 feet long. The pressure is 80 psig at the top and 85 psig at the bottom. A ball type check valve is installed near the bottom. The tube is steel having a $1\frac{1}{4}$-inch outside diameter and a 0.109-inch wall thickness. Compute the volume flow rate of water.

9–12   Turpentine at 77°F is flowing from A to B in a 3-inch asphalt-coated cast iron pipe (actual inside diameter = 3.58 inches). Point B is 20 feet higher than A and the total length of pipe is 60 feet. Two 90° long radius elbows are installed between A and B. Calculate the volume flow rate of turpentine if the pressure at A is 120 psig and the pressure at B is 105 psig.

9–13   A device designed to allow cleaning of walls and windows on the second floor of homes is similar to the system shown in Fig. 9–15. Determine the velocity of flow from the nozzle if the pressure at the bottom is (a) 20 psig, (b) 80 psig. The nozzle has a loss coefficient C_L of 0.15 based on the outlet velocity. The tube is smooth drawn aluminum having an inside diameter of 0.5 inches. The 90° bend has a radius of 6 inches. The total length of straight tube is 20 feet. The fluid is water at 100° F.

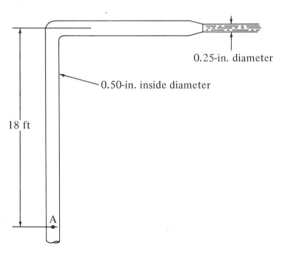

0.25-in. diameter

0.50-in. inside diameter

18 ft

A

FIGURE 9–15

9–14   Kerosene at 77°F is flowing in the system shown in Fig. 9–16. The total length of 2-inch Type K copper tubing is 100 feet. The two 90° bends have a radius of 12 inches. Calculate the volume flow rate into tank B if a pressure of 20 psig is maintained above the kerosene in tank A.

9–15   Water at 100°F is flowing from A to B through the system shown in Fig. 9–17. Determine the volume flow rate of water if the vertical distance between the surfaces of the two reservoirs is 30 feet. Both pipes are asphalt coated cast iron. The elbows are standard.

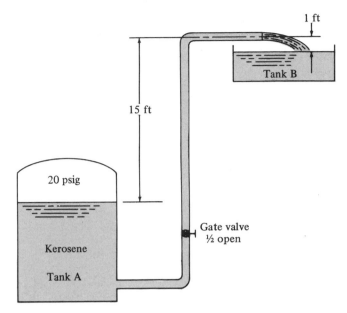

FIGURE 9–16

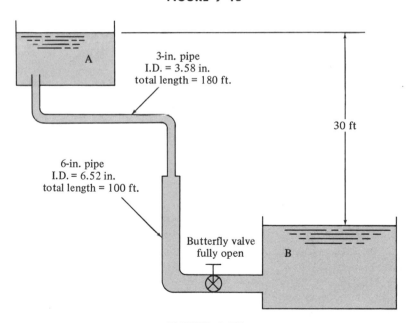

FIGURE 9–17

9–16 Oil having a specific gravity of 0.93 and a dynamic viscosity of 2.0×10^{-4} lb-sec/ft² is flowing into the open tank shown in

Fig. 9–18. The total length of 2-inch tubing is 100 feet and of
4-inch tubing is 300 feet. The elbows are standard. Determine the
volume flow rate into the tank if the pressure at point *A* is 25 psig.

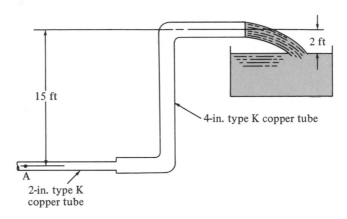

FIGURE 9–18

9–17   Determine the required size of new Schedule 80 steel pipe to
carry water at 160°F with a maximum pressure drop of 10 psi
per 1000 feet when the flow rate is 0.5 ft³/sec.

9–18   What size of standard Type L copper tube is required to transfer
2.0 ft³/sec of water at 160°F from a heater where the pressure
is 20 psig to an open tank? The tube is horizontal and 100 feet
long.

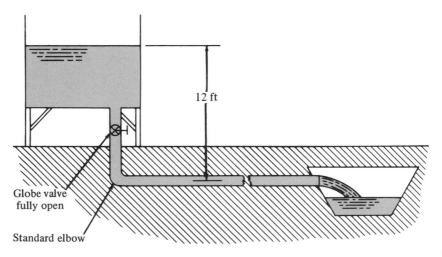

FIGURE 9–19

9–19   Water at 60°F is to flow by gravity between two points, two miles apart, at the rate of 13,500 gal/min. The upper end is 130 feet higher than the lower end. What size concrete pipe is required?

9–20   The tank shown in Fig. 9–19 is to be drained to a sewer. Determine the size of new Schedule 40 steel pipe which will carry at least 400 gal/min of water at 80°F through the system shown. The total length of pipe is 75 feet.

10

Parallel Pipe Line Systems

10–1   Differences Between Series and Parallel Systems

If a pipe line system is arranged so that the fluid flows in a continuous line without branching, it is referred to as a series system. Conversely, if the system causes the flow to branch into two or more lines it is referred to as a parallel system.

The nature of parallel systems requires that the approach to their analysis be different from that used for series systems. In general, a parallel system can have any number of branches. The system shown in Fig. 10–1, having three branches, will be used to illustrate the basic concepts. The flow in the main line at section one splits into three parts and then rejoins at section two. For most problems of this type, the objective is to determine how much of the flow occurs in each of the branches, and how much pressure drop occurs between sections 1 and 2.

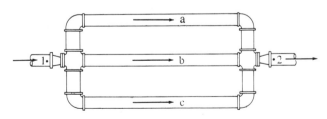

FIGURE 10–1

The following equations state the principles on which the solution procedures for parallel systems are based.

$$Q_1 = Q_2 = Q_a + Q_b + Q_c \qquad \text{(10–1)}$$
$$h_{L_{1-2}} = h_a = h_b = h_c \qquad \text{(10–2)}$$

Equation (10–1) is the statement of the condition of continuity for steady flow in a parallel system. In Eq. (10–2), the term $h_{L_{1-2}}$ is the energy loss *per pound of fluid* between the two points 1 and 2 in the main lines. The terms h_a, h_b, and h_c are the energy losses *per pound of fluid* in each branch of the system. The fact that these quantities must all be equal can be demonstrated by writing the energy equation using points 1 and 2 as the reference points.

$$\frac{p_1}{\gamma} + z_1 + \frac{v_1{}^2}{2g} - h_L = \frac{p_2}{\gamma} + z_2 + \frac{v_2{}^2}{2g} \qquad (10\text{–}3)$$

In earlier chapters, the sum of the pressure head p/γ, the elevation head z_1, and the velocity head $v^2/2g$ was referred to as the total head E. This represents the energy possessed by each pound of the fluid at a particular point in a system. Substituting E into Eq. (10–3) gives

$$E_1 - h_L = E_2$$

and

$$h_L = E_1 - E_2 \qquad (10\text{–}4)$$

Then the term h_L represents the loss of head between points 1 and 2. In Fig. 10–1, each pound of fluid has the same total head at the point where the flow branches. As the fluid flows through the branches, some energy is lost. But at the point where the flow rejoins, the total head of each pound of fluid must again be the same. The conclusion can thus be drawn that the loss of head is the same, regardless of which path is taken between points 1 and 2. This conclusion is stated mathematically as Eq. (10–2).

The amount of fluid flowing in a particular branch of a parallel system is dependent on the resistance to flow in that branch relative to the resistance in other branches. The fluid will tend to follow the path of least resistance. As stated in Chapters 7 and 8, resistance to flow is due to pipe wall friction, to changes in the cross section of the flow path, to changes in the direction of flow, or to obstructions such as might exist in valves. All of these resistances are dependent on the velocity of flow. Therefore, in a parallel system, the flow divides in such a way that the velocities are different in the various branches so that the loss of head in each branch is equal.

A common parallel piping system includes two branches arranged as shown in Fig. 10–2. The lower branch is added to allow some fluid to bypass the heat exchanger. The branch could also be used to isolate the heat exchanger, allowing continuous flow while the equipment is serviced. The analysis of this type of system is relatively simple and straightforward as will be explained in the following section.

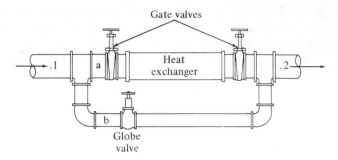

FIGURE 10–2

Parallel systems having more than two branches are more complex because there are more unknown quantities than there are equations relating the unknowns. A solution procedure requiring iteration is described in Section 10–3.

10–2   **Systems with Two Branches**

The system shown in Fig. 10–2 will be used to illustrate the analysis of flow in two branches. The basic relationships which apply here are similar to Eqs. (10–1) and (10–2) except there are only two branches instead of three.

$$Q_1 = Q_2 = Q_a + Q_b \tag{10-5}$$
$$h_{L_{1-2}} = h_a = h_b \tag{10-6}$$

The following example problem is presented in the programmed form. Careful attention should be given to the logic of the solution procedure as well as to the details performed.

Programmed Example Problem

Example Problem 10–1. In Fig. 10–2, 100 gal/min of water at 60°F is flowing in a 2-inch Schedule 40 steel pipe at section 1. The heat exchanger in branch a has a loss coefficient of $C_L = 12.5$. All three valves are wide open. Branch b is a bypass line composed of $1\frac{1}{4}$-inch Schedule 40 steel pipe. The elbows are standard. The length of pipe between points 1 and 2 in branch b is 20 feet. Because of the size of the heat exchanger the length of pipe in branch a is very short and friction losses can be neglected. For this arrangement, determine the volume flow rate of water in each branch and the pressure drop between points 1 and 2.

Equation (10–5) relates the two volume flow rates. How many quantities are unknown in this equation?

The two velocities, v_a and v_b are unknown. Since $Q = Av$, Eq. (10–5) can be expressed as

$$Q_1 = A_a v_a + A_b v_b \qquad (10\text{–}7)$$

From the given data, $A_a = 0.02333$ ft², $A_b = 0.01039$ ft², and $Q_1 = 100$ gal/min. Expressing Q_1 in the units of ft³/sec gives

$$Q_1 = 100 \text{ gal/min} \times \frac{1 \text{ ft}^3/\text{sec}}{449 \text{ gal/min}} = 0.223 \text{ ft}^3/\text{sec}$$

Another equation must be generated which also relates v_a and v_b.

Equation (10–6) states that the head loss in each branch is equal. Since the head losses h_a and h_b are dependent on the velocities v_a and v_b, this equation can be used in conjunction with Eq. (10–7) to solve for the velocities. Now, express the head losses in terms of the velocities for each branch.

You should have something similar to this.

$$h_a = 2 f_a (L_e/D)(v_a^2/2g) + 12.5(v_a^2/2g)$$
$$\text{(valves)} \qquad\qquad \text{(heat exchanger)}$$

Using $L_e/D = 13$ for the fully open gate valves,

$$h_a = (26 f_a + 12.5)(v_a^2/2g) \qquad (10\text{–}8)$$

Also,

$$h_b = 2 f_b(L_e/D)(v_b^2/2g) + f_b(L_e/D)(v_b^2/2g) + f_b(L/D)(v_b^2/2g)$$
$$\text{(elbows)} \qquad\qquad \text{(valve)} \qquad\qquad \text{(friction)}$$

Using $L_e/D = 30$ for the elbows, $L_e/D = 340$ for the globe valve, $L = 20$ ft, and $D = 0.1150$ ft,

$$h_b = 574 f_b(v_b^2/2g) \qquad (10\text{–}9)$$

Equations (10–8) and (10–9) have introduced two additional unknowns, the friction factors f_a and f_b. The problem now appears to be similar to the Class II systems discussed in Chapter 9. If values for the friction factors can be assumed, then the velocities can be calculated. The friction factors can then be reevaluated using an iteration technique if necessary.

The values of the relative roughness, D/ϵ, can be used to determine initial estimates for f_a and f_b.

$$(D/\epsilon)_a = (0.1723/1.5 \times 10^{-4}) = 1150$$
$$(D/\epsilon)_b = (0.1150/1.5 \times 10^{-4}) = 768$$

From the Moody diagram, Fig. 7–2, logical estimates for the friction factors are $f_a = 0.020$ and $f_b = 0.023$. Substitute these values into Eqs. (10–8) and (10–9) now. Then, using the fact that $h_a = h_b$, solve for v_a in terms of v_b.

You should have $v_a = 1.007\, v_b$. Here is how it is done.

$$h_a = (26\,f_a + 12.5)(v_a{}^2/2g) \qquad\qquad \text{(10–8)}$$
$$= [(26)(0.020) + 12.5](v_a{}^2/2g) = 13.02\,(v_a{}^2/2g)$$
$$h_b = 574\,f_b\,(v_b{}^2/2g) \qquad\qquad\qquad\quad \text{(10–9)}$$
$$= 574\,(0.023)(v_b{}^2/2g) = 13.20\,(v_b{}^2/2g)$$

But $h_a = h_b$. Then,

$$13.02\,(v_a{}^2/2g) = 13.20\,(v_b{}^2/2g)$$
$$v_a = 1.007\,v_b \qquad\qquad \text{(10–10)}$$

What should be the next step in the solution of this problem?

At this time, Eqs. (10–7) and (10–10) can be combined to calculate the velocities.

The solutions are $v_a = 6.62$ ft/sec and $v_b = 6.58$ ft/sec. Here are the details.

$$Q_1 = A_a v_a + A_b v_b \qquad\qquad \text{(10–7)}$$
$$v_a = 1.007\,v_b \qquad\qquad\qquad \text{(10–10)}$$

Then,

$$Q_1 = A_a(1.007\,v_b) + A_b v_b = v_b(1.007\,A_a + A_b)$$

Solving for v_b.

$$v_b = \frac{Q_1}{1.007\,A_a + A_b} = \frac{0.223 \text{ ft}^3/\text{sec}}{[(1.007)(0.02333) + 0.01039] \text{ ft}^2}$$
$$v_b = 6.58 \text{ ft/sec}$$
$$v_a = (1.007)(6.58) \text{ ft/sec} = 6.62 \text{ ft/sec}$$

It is quite coincidental that the two velocities are so close. Since these calculations were made using assumed values for f_a and f_b, it is necessary to check the accuracy of the assumptions. By evaluating the Reynolds numbers and referring to the Moody diagram, we find $f_a = 0.021$ and $f_b = 0.024$. These are so close to the originally assumed values that no significant changes in the velocities would result from repeating the calculations.

Now calculate the volume flow rates Q_a and Q_b.

You should have

$$Q_a = A_a v_a = (0.02333 \text{ ft}^2)(6.62 \text{ ft/sec}) = 0.1545 \text{ ft}^3/\text{sec}$$
$$Q_b = A_b v_b = (0.01039 \text{ ft}^2)(6.58 \text{ ft/sec}) = 0.0683 \text{ ft}^3/\text{sec}$$

Converting these values to the units of gal/min gives $Q_a = 69.4$ gal/min and $Q_b = 30.6$ gal/min.

The problem statement also asked that the pressure drop be calculated. How can this be done?

The energy equation can be written between points 1 and 2. Since the velocities and elevations are the same at these points, the energy equation is simply

$$\frac{p_1}{\gamma} - h_L = \frac{p_2}{\gamma}$$

Solving for the pressure drop,

$$p_1 - p_2 = \gamma h_L \qquad \qquad \textbf{(10–11)}$$

What can be used to calculate h_L?

Since $h_{L_{1-2}} = h_a = h_b$, either Eq. (10–8) or (10–9) can be used. Using Eq. (10–8),

$$h_a = 13.02 \ (v_a{}^2/2g) = (13.02)(6.62)^2/64.4 \text{ ft} = 8.90 \text{ ft}$$

Then,

$$p_1 - p_2 = \gamma h_L = \frac{62.4 \text{ lb}}{\text{ft}^3} \cdot 8.90 \text{ ft} \cdot \frac{1 \text{ ft}^2}{144 \text{ in.}^2} = 3.85 \text{ psi}$$

This example problem is concluded.

Another programmed example problem follows which illustrates a different type of branched system. The pressure drop across the parallel branches is being measured and the volume flow rate is to be determined.

Example Problem 10–2. The arrangement shown in Fig. 10–3 is being used to supply lubricating oil to the bearings of a large machine. The bearings act as restrictions to the flow. The loss coefficients are 11.0 and 4.0 for the two bearings. The lines in each branch are $\frac{1}{2}$-inch steel tubing having a wall thickness of 0.049 inch. Each of the four bends in the tubing has a radius of 4 inches. Include the effect of these bends but exclude the friction losses since the lines are short. Determine the flow rate of oil in each bearing and the total flow rate in gallons per minute. The oil has a specific gravity of 0.881 and a kinematic viscosity of 2.50 $\times$ 10^{-5} ft^2/sec.

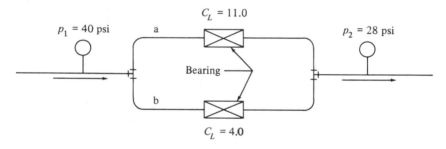

FIGURE 10–3

Write the equation which relates the head loss h_L across the parallel system to the head losses in each line h_a and h_b.

You should have

$$h_L = h_a = h_b \qquad\qquad\qquad (10\text{–}12)$$

They are all equal. Can we determine the magnitude of these head losses?

Yes. From the energy equation we can find h_L.

$$\frac{p_1}{\gamma} + z_1 + \frac{v_1^2}{2g} - h_L = \frac{p_2}{\gamma} + z_2 + \frac{v_2^2}{2g}$$

But $z_1 = z_2$ and $v_1 = v_2$. Then,

$$\frac{p_2}{\gamma} = \frac{p_1}{\gamma} - h_L$$

$$h_L = (p_1 - p_2)/\gamma \qquad (10\text{–}13)$$

Using the given data,

$$h_L = \frac{(40 - 28)\ \text{lb}}{\text{in.}^2} \cdot \frac{\text{ft}^3}{(0.881)(62.4)\ \text{lb}} \cdot \frac{144\ \text{in.}^2}{\text{ft}^2}$$

$$h_L = 31.4\ \text{ft}$$

Now, write the expressions for h_a and h_b.

Considering the losses in the bends and in the bearings you should have

$$h_a = 2f\,(L_e/D)\,v_a{}^2/2g + 11.0\ v_a{}^2/2g \qquad (10\text{–}14)$$
$$h_b = 2f\,(L_e/D)\,v_b{}^2/2g + 4.0\ v_b{}^2/2g \qquad (10\text{–}15)$$

The equivalent length ratio for the bends can be found from Fig. 8–19 in Chapter 8. For each bend the relative radius is

$$r/D = 4\ \text{in.}/0.402\ \text{in.} = 9.95$$

Then $L_e/D = 29.5$. The friction factor can be estimated using the relative roughness as a guide.

$$D/\epsilon = 0.0335\ \text{ft}/1.5 \times 10^{-4}\ \text{ft} = 223$$

From the Moody diagram we can estimate the friction factor to be 0.032. Then,

$$h_a = (2)(0.032)(29.5)(v_a{}^2/2g) + 11.0\ (v_a{}^2/2g)$$
$$h_a = (1.89 + 11.0)\ v_a{}^2/2g = 12.89\ v_a{}^2/2g \qquad (10\text{–}16)$$
$$h_b = (2)(0.032)(29.5)(v_b{}^2/2g) + 4.0\ (v_b{}^2/2g)$$
$$h_b = (1.89 + 4.0)\ v_b{}^2/2g = 5.89\ v_b{}^2/2g \qquad (10\text{–}17)$$

Can the velocities v_a and v_b be found now?

Yes, they can. It was found earlier that $h_L = 31.4$ ft. Since $h_L = h_a = h_b$, Eqs. (10–16) and (10–17) can be solved directly for v_a and v_b.

$$h_a = 12.89\ v_a{}^2/2g$$

$$v_a = \sqrt{\frac{2gh_a}{12.89}} = \sqrt{\frac{(2)(32.2)(31.4)}{12.89}}\ \text{ft/sec}$$

$$v_a = 12.52\ \text{ft/sec}$$

$$h_b = 5.89\ v_b{}^2/2g$$

$$v_b = \sqrt{\frac{2gh_b}{5.89}} = \sqrt{\frac{(2)(32.2)(31.4)}{5.89}}\ \text{ft/sec}$$

$$v_b = 18.5\ \text{ft/sec}$$

The assumed value for the friction factor was used in obtaining this result. A check of the Reynolds numbers gives

$$N_{R_a} = \frac{v_a D}{\nu} = \frac{(12.52)(0.0335)}{2.50 \times 10^{-5}} = 1.68 \times 10^4$$

$$N_{R_b} = \frac{v_b D}{\nu} = 2.48 \times 10^4$$

Using $D/\epsilon = 223$ for both tubes gives a value for f very close to the assumed value of 0.032. Therefore, the velocities found are correct. Now find the volume flow rates.

You should have $Q_a = 4.96$ gal/min, $Q_b = 7.34$ gal/min, and the total volume flow rate = 12.30 gal/min. The area of each tube is 8.814 $\times 10^{-4}$ ft^2. Then,

$$Q_a = A_a v_a = 8.814 \times 10^{-4} \text{ ft}^2 \cdot 12.52 \text{ ft/sec} \cdot \frac{449 \text{ gal/min}}{1 \text{ ft}^3/\text{sec}}$$

$$Q_a = 4.96 \text{ gal/min}$$

Similarly,

$$Q_b = A_b v_b = 7.34 \text{ gal/min}$$

Then the total flow rate is

$$Q = Q_a + Q_b = (4.96 + 7.34) \text{ gal/min} = 12.30 \text{ gal/min}$$

This example problem is concluded.

10–3   Systems with Three or More Branches

When three or more branches occur in a pipe flow system, it is called a network. Networks are indeterminate since there are more unknown factors than there are independent equations relating the factors. For example, in Fig. 10–4 there are three velocities unknown, one in each pipe. The equations available to describe the system are

$$Q_1 = Q_2 = Q_a + Q_b + Q_c \qquad (10\text{–}18)$$
$$h_{L_{1-2}} = h_a = h_b = h_c \qquad (10\text{–}19)$$

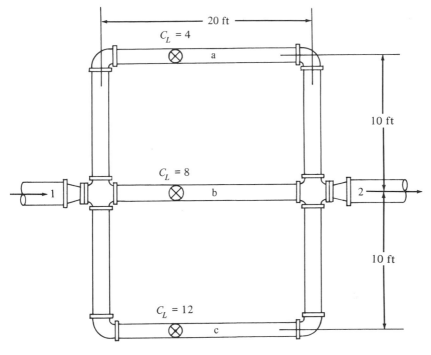

FIGURE 10–4

Note: Inlet and outlet pipes: 2-in. Sch. 40
Branch pipes a, b, and c: 1-in. Sch. 40
Elbows are standard

A third independent equation is required to solve explicitly for the three velocities, and none is available.

It would be possible to complete the analysis of such a system as that shown in Fig. 10–4 by a trial and error technique, assuming different values for the velocities until Eqs. (10–18) and (10–19) are satisfied. However, many calculations would be required. Also, if the system is composed of additional pipes, the complexity increases dramatically.

A more rational approach employing an iteration procedure has been developed by Hardy Cross.* This procedure converges on the correct flow rates quite rapidly. A significant number of calculations are still required but they can be set up in an orderly fashion for use on a calculator or digital computer.

Hardy Cross. *Analysis of Flow in Networks of Conduits or Conductors.* University of Illinois Engineering Experiment Station Bulletin No. 286 (Urbana: University of Illinois, November, 1936).

The Cross technique requires that the head loss terms for each pipe in the system be expressed in the form

$$h = kQ^n \qquad (10\text{--}20)$$

where k is an equivalent resistance to flow for the entire pipe and Q is the flow rate in the pipe. To illustrate the procedure for determining the value of k, consider branch a of the system in Fig. 10–4. The total head loss is due to the two elbows, the restriction having a loss coefficient of 4.0, and friction in the pipe. Then, using $L_e/D = 30$ for standard elbows,

$$h_a = \underset{\text{(elbows)}}{2f\,(30)\,v_a^2/2g} + \underset{\text{(restriction)}}{4.0\,v_a^2/2g} + \underset{\text{(friction)}}{f\,(40/0.0874)\,v_a^2/2g}$$

It is necessary to assume a value for the friction factor. In order to simplify the remainder of the procedure, it will be assumed that the friction factor is constant. In reality, of course, its value changes for different flow rates, and this would have to be taken into account to obtain very accurate solutions. A reasonable estimate for f can be made by first calculating the relative roughness, D/ϵ.

$$D/\epsilon = 0.0874/1.5 \times 10^{-4} = 583$$

From the Moody Diagram, $f = 0.025$ would be a fair estimate. Then the head loss equation becomes

$$h_a = (1.5 + 4.0 + 11.45)\,v_a^2/2g = 16.95\,v_a^2/2g$$

In order to put this into the form of Eq. (10–20), the velocity must be expressed in terms of the volume flow rate.

$$v_a^2 = Q_a^2/A^2$$

Then,

$$h_a = \frac{16.95\,Q_a^2}{2g\,A^2} = \frac{16.95\,Q_a^2}{(2)(32.2)(0.0060)^2}$$

$$h_a = 7300\,Q_a^2 \qquad (10\text{--}21)$$

Now the expression for h_a is in the required form, where $k_a = 7300$ and $n = 2$. Similar calculations must be made for h_b and h_c.

Since iteration is to be used to complete the analysis of the network, an initial trial value must be assumed for the volume flow rate Q in each pipe. Some factors which help in making these estimates are:

1. At each junction in the network, the sum of the flow rates into the junction must equal the flow out.
2. The fluid tends to follow the path of least resistance through the network. Pipes having a high value of k will have relatively low flow rates.

In the solution procedure outlined below, the circuit is divided into several closed loop circuits. An adjustment to the assumed value of Q is calculated for each circuit. The adjustment, ΔQ, is algebraically subtracted from the flow rate in each pipe. The sign convention to be used for the circuits is illustrated in Fig. 10–5, which is a schematic

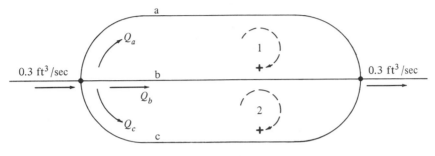

FIGURE 10–5

representation of the network in Fig. 10–4. There are two closed loop circuits indicated by the dashed arrows labeled 1 and 2. The flow in each pipe is assumed to be in the direction shown. The sign convention is:

If the flow is clockwise, Q and h are positive.
If the flow is counterclockwise, Q and h are negative.

Then for circuit 1 in Fig. 10–5, h_a and Q_a are positive while h_b and Q_b are negative. In circuit 2, h_b and Q_b are positive while h_c and Q_c are negative. Notice that pipe b is common to both circuits. Therefore, the adjustment, ΔQ, for each circuit must be applied to the flow rate in this pipe.

The Cross technique for analyzing the flow in pipe networks is presented in step-by-step form below. A programmed example problem follows to illustrate the application of the procedure.

Cross Technique for Analysis of Pipe Networks

1. Express the energy loss in each pipe in the form, $h = kQ^2$.
2. Assume a value for the flow rate in each pipe such that the flow into each junction equals the flow out of the junction.
3. Divide the network into a series of closed loop circuits.
4. For each pipe, calculate the head loss, $h = kQ^2$, using the assumed value of Q.
5. Proceeding around each circuit in a clockwise direction, algebraically sum all values for h using the following sign convention.

If the flow is clockwise, h *and* Q *are positive.*
If the flow is counterclockwise, h *and* Q *are negative.*
The resulting summation is referred to as Σh.

6. For each pipe, calculate $2kQ$.
7. Sum all values of $2kQ$ for each circuit, assuming all are positive. This summation is referred to as $\Sigma(2kQ)$.
8. For each circuit, calculate the value of ΔQ from

$$\Delta Q = \frac{\Sigma h}{\Sigma(2kQ)} \qquad (10\text{--}22)$$

9. For each pipe, calculate a new estimate for Q from

$$Q' = Q - \Delta Q$$

10. Repeat Steps 4–8 until ΔQ from Step 8 becomes negligibly small. The Q' value is used for the next cycle of iteration.

Programmed Example Problem

Example Problem 10–3. For the system shown in Fig. 10–4 determine the volume flow rate of water through each branch if 0.30 ft³/sec is flowing into and out of the system through the two inch pipes.

The energy loss in each pipe should now be expressed in the form $h = kQ^2$ in order to complete Step 1 of the procedure. Equation (10–21) gives $h_a = 7300 \, Q_a{}^2$ for pipe *a*. Do likewise for h_b and h_c before looking at the next panel.

You should have $h_b = 5930 \, Q_b{}^2$ and $h_c = 10{,}730 \, Q_c{}^2$. Here are the details. For pipe *b*,

$$h_b = 8.0 \, v_b{}^2/2g + f \, (20/0.0874) \, v_b{}^2/2g$$
$$\text{(restriction)} \quad \text{(friction)}$$

Since $D/\epsilon = 583, f = 0.025$ is assumed. Then

$$h_b = (8.0 + 5.73) \, v_b{}^2/2g = 13.73 \, v_b{}^2/2g$$

But $v_b{}^2 = Q_b{}^2/A^2$.

$$h_b = \frac{13.73 \, Q_b{}^2}{2g \, A^2} = \frac{13.73 \, Q_b{}^2}{(2)(32.2)(0.006)^2} = 5930 \, Q_b{}^2$$

For pipe *c*,

$$h_c = 2f \, (L_e/D) \, v_c{}^2/2g + 12.0 \, v_c{}^2/2g + f \, (40/0.0874) \, v_c{}^2/2g$$
$$\text{(elbows)} \qquad \text{(restriction)} \qquad \text{(friction)}$$

Using $f = 0.025$ and $L_e/D = 30$,

$$h_c = (1.5 + 12.0 + 11.45) \, v_c{}^2/2g = 24.95 \, v_c{}^2/2g$$
$$h_c = \frac{24.95 \, Q_c{}^2}{2g \, A^2} = \frac{24.95 \, Q_c{}^2}{(2)(32.2)(0.006)^2} = 10{,}730 \, Q_c{}^2$$

Step 2 of the procedure asks that estimates be given for the volume flow rate in each branch. Which pipe should have the greatest flow rate, and which should have the least?

Pipe b has the least value for k and therefore it should carry the greatest flow. Conversely, branch c has the highest value for k. Then Q_c should be the smallest of the three. Many different first estimates are possible for the flow rates. However, it is known that

$$Q_a + Q_b + Q_c = 0.3 \text{ ft}^3/\text{sec}$$

Assume $Q_a = 0.10 \text{ ft}^3/\text{sec}$, $Q_b = 0.11 \text{ ft}^3/\text{sec}$, and $Q_c = 0.09 \text{ ft}^3/\text{sec}$ as a start.

For Step 3 of the procedure, the circuits are shown in Fig. 10–5. Now complete Step 4.

Check your results with these.

$$h_a = k_a Q_a^2 = (7300)(0.10)^2 = 73.0$$
$$h_b = k_b Q_b^2 = (5930)(0.11)^2 = 71.8$$
$$h_c = k_c Q_c^2 = (10,730)(0.09)^2 = 87.0$$

Now do Step 5.

For circuit 1,

$$\Sigma h_1 = h_a - h_b = 73.0 - 71.8 = +1.2$$

For circuit 2,

$$\Sigma h_2 = h_b - h_c = 71.8 - 87.0 = -15.2$$

Now do Step 6.

Here are the correct values.

Pipe a: $2k_a Q_a = (2)(7300)(0.10) = 1460$
Pipe b: $2k_b Q_b = (2)(5930)(0.11) = 1302$
Pipe c: $2k_c Q_c = (2)(10,730)(0.09) = 1935$

Now do Step 7.

For circuit 1,

$$\Sigma(2kQ)_1 = 1460 + 1302 = 2762$$

For circuit 2,

$$\Sigma(2kQ)_2 = 1302 + 1935 = 3237$$

The adjustment for the flow rates, ΔQ, can now be calculated for each circuit using Step 8.

For circuit 1,

$$\Delta Q_1 = \frac{\Sigma h_1}{\Sigma (2kQ)_1} = \frac{1.2}{2762} = 0.00043$$

For circuit 2,

$$\Delta Q_2 = \frac{\Sigma h_2}{\Sigma (2kQ)_2} = \frac{-15.2}{3237} = -0.00470$$

These values for ΔQ are estimates of the error in the originally assumed values for Q. It is recommended that the process be repeated until the magnitude of ΔQ is less than 1 percent of the assumed value of Q. Special circumstances may warrant using a different criterion for judging ΔQ.

Step 9 can now be completed. Calculate the new value for Q_a before looking at the next panel.

The calculation is as follows:

$$Q_a' = Q_a - \Delta Q_1 = 0.10 - 0.00043$$
$$Q_a' = 0.09957 \text{ ft}^3/\text{sec}$$

Calculate the new value for Q_c before Q_b. Pay careful attention to algebraic signs.

You should have

$$Q_c' = Q_c - \Delta Q_2 = -0.09 - (-0.00470)$$
$$Q_c' = -0.08530 \text{ ft}^3/\text{sec}$$

Notice that Q_c is negative since it flows in a counterclockwise direction in circuit 2. The calculation for Q_c' can be interpreted as indicating that the magnitude of Q_c must be decreased in absolute value.

Now calculate the new value for Q_b. Remember, pipe b is in each circuit.

Both ΔQ_1 and ΔQ_2 must be applied to Q_b. For circuit 1,

$$Q_b' = Q_b - \Delta Q_1 = -0.11 - 0.00043$$

This would result in an increase in the absolute value of Q_b. For circuit 2,

$$Q_b' = Q_b - \Delta Q_2 = +0.11 - (-0.00470)$$

This also results in increasing Q_b. Then Q_b is actually increased by the sum of ΔQ_1 and ΔQ_2 in absolute value. That is,

$$Q_b' = 0.11 + 0.00043 + 0.00470$$
$$Q_b' = 0.11513 \text{ ft}^3/\text{sec}$$

Remember that the sum of the absolute values of the flow rates in the three pipes must equal 0.30 ft³/sec, the total Q.

The iteration can be continued by using Q_a', Q_b', and Q_c' as the new estimates for the flow rates and repeating Steps 4–8. The results for four iteration cycles are summarized in the table presented in the next panel. You should carry out the calculations yourself.

TABLE 10–1

Trial	Circuit	Pipe	Q	$h = kQ^2$	$2\,kQ$	ΔQ
1	1	a	0.100	73.0	1460	
		b	−0.110	−71.8	1302	
				+ 1.2	2762	+0.00043
	2	b	0.110	71.8	1302	
		c	−0.090	−87.0	1935	
				−15.2	3237	−0.00470
2	1	a	0.09957	72.5	1455	
		b	−0.11513	−78.6	1367	
				−6.1	2822	−0.00216
	2	b	0.11513	78.6	1367	
		c	−0.08530	−78.0	1830	
				+0.6	3197	+0.00019
3	1	a	0.10173	75.4	1481	
		b	−0.11279	−75.2	1335	
				+0.2	2816	+0.00007
	2	b	0.11279	75.2	1335	
		c	−0.08549	−78.4	1837	
				−3.2	3172	−0.00101
4	1	a	0.10166	75.3	1482	
		b	−0.11387	−76.6	1348	
				−1.3	2830	−0.00046
	2	b	0.11387	76.6	1348	
		c	−0.08448	−76.5	1812	
				+0.1	3160	+0.00003

Note: $k_a = 7300$, $k_b = 5930$, $k_c = 10{,}730$

The results show that $Q_a = 0.10166$ ft³/sec, $Q_b = 0.11387$ ft³/sec, and $Q_c = 0.08448$ ft³/sec in the directions shown in Fig. 10–5. Converted to gal/min, these values are $Q_a = 45.6$ gal/min, $Q_b = 51.0$ gal/min, and $Q_c = 37.8$ gal/min.

PRACTICE PROBLEMS

10–1 Figure 10–6 shows a branched system in which the pressure at A is 100 psig and the pressure at B is 80 psig. Each branch is 200 feet long. Use a friction factor of 0.025 as a first estimate for both pipes. Neglect losses at the junctions but consider all elbows. If the system carries oil with a specific weight of 56.0 lb/ft³, calculate the total volume flow rate. The oil has a kinematic viscosity of 5.2×10^{-5} ft²/sec.

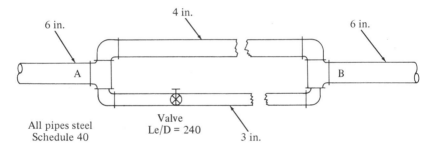

FIGURE 10–6

10–2 Using the system shown in Fig. 10–2 and the data from Example Problem 10–1 determine the volume flow rate of water in each branch and the pressure drop between points 1 and 2 if the first gate valve is one half closed and the other valves are wide open.

10–3 In the branched pipe system shown in Fig. 10–7, 0.5 ft³/sec of water at 60°F is flowing in a 4-inch Schedule 40 pipe at A. The flow splits into two 2-inch Schedule 40 pipes as shown and then rejoins at B. Calculate the flow rate in each of the branches and the pressure difference $p_A - p_B$. Include the effect of the minor losses in the lower branch of the system. Use $f = 0.03$ for both pipes. The total length of pipe in the lower branch is 200 feet. The elbows are standard.

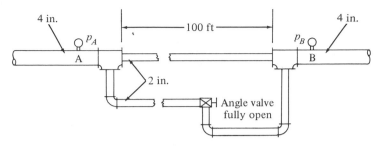

FIGURE 10-7

10-4 In the branched pipe system shown in Fig. 10–8, 1350 gal/min of benzene (sg = 0.87) at 140°F is flowing in the 8-inch pipe. Calculate the volume flow rate in the 6-inch and the 2-inch pipes. All pipes are standard Schedule 40 steel pipes. Use $f = 0.02$ for a first estimate and refine the choice of f after the first calculations. Neglect the elbows.

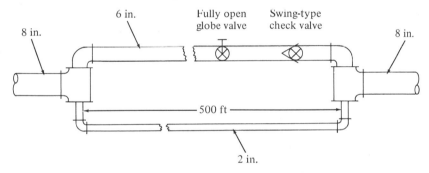

FIGURE 10-8

10-5 A 6-inch pipe branches into a 4-inch and a 2-inch pipe as shown in Fig. 10–9. Both pipes are 100 feet long and have a friction factor of 0.02. Neglect the losses in the elbows. Determine

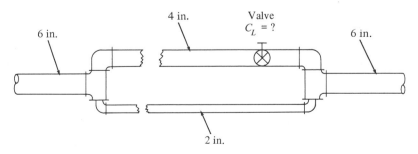

FIGURE 10-9

what the loss coefficient C_L of the valve must be in order to obtain equal volume flow rates in each branch.

10–6   For the system shown in Fig. 10–10, the pressure at A is maintained constant at 20 psig. The total volume flow rate exiting from the pipe at B depends on which valves are open or closed.

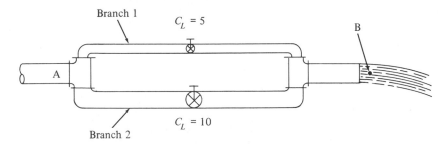

FIGURE 10–10

Use $C_L = 0.9$ for each elbow but neglect the energy losses in the tees. Also, since the length of each branch is short, neglect pipe friction losses. The pipe in branch 1 has a 2-inch inside diameter and branch 2 has a 4-inch inside diameter. Calculate the volume flow rate of water for each of the following conditions:

    a.   Both valves open
    b.   Valve in branch 2 only open
    c.   Valve in branch 1 only open

10–7   Solve problem 10–4 using the Hardy Cross technique.

10–8   Solve problem 10–3 using the Hardy Cross technique.

10–9   Find the flow rate in each pipe of Fig. 10–11. Use $f = 0.015$ for all pipes.

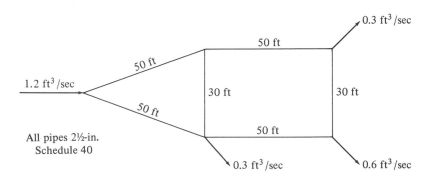

FIGURE 10–11

10–10 Figure 10–12 represents a spray rinse system in which water at 60°F is flowing. All pipes are 3-inch Type L copper tubing. Assume $f = 0.015$ for all pipes. Determine the flow rate in each pipe.

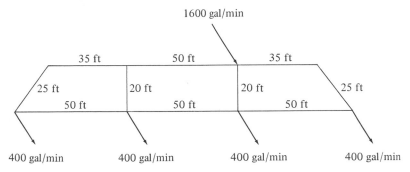

FIGURE 10–12

10–11 Figure 10–13 represents the water distribution network in a small industrial park. The supply of 15.5 ft³/sec enters the system at A. Manufacturing plants draw off the indicated flows at points C, E, F, G, H, and I. Determine the flow in each pipe in the system. Use $f = 0.02$ for all pipes.

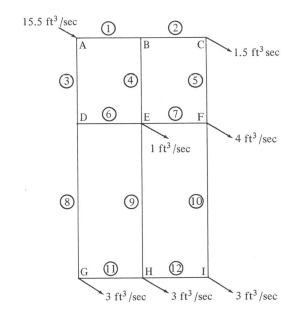

Pipe Data		
All pipes Schedule 40		
Pipe No.	Length (ft)	Size (in.)
1	1500	16
2	1500	16
3	2000	18
4	2000	12
5	2000	16
6	1500	16
7	1500	12
8	4000	14
9	4000	12
10	4000	8
11	1500	12
12	1500	8

FIGURE 10–13

10–12   Figure 10–14 represents the network for delivering coolant to five different machine tools in an automated machining system. The grid is a rectangle 25 feet by 50 feet. All pipes are commercial steel tubing having a 0.065-inch wall thickness. Pipes 1 and 3 are 2-inch diameter; pipe 2 is 1½-inch diameter; and all others are 1-inch. Use $f = 0.015$ for all pipes.

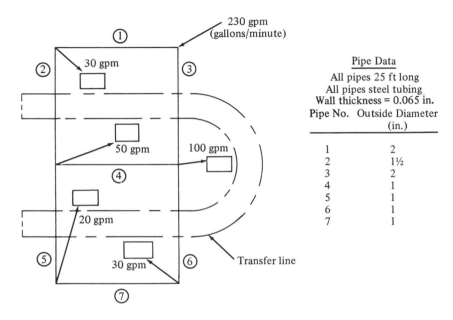

Pipe Data

All pipes 25 ft long
All pipes steel tubing
Wall thickness = 0.065 in.

Pipe No.	Outside Diameter (in.)
1	2
2	1½
3	2
4	1
5	1
6	1
7	1

FIGURE 10–14

11

Flow in Noncircular Cross Sections and Open Channel Flow

Introduction

Noncircular cross sections can be either closed conduits or open channels and the overall analysis of each type is quite different from the other. However, some similarities exist, and it is convenient to relate the principles which govern the flow in each type of noncircular section to that in circular sections, to which much effort has already been devoted.

Examples of typical noncircular cross sections are shown in Fig. 11–1 for closed conduits and in Fig. 11–2 for open channels. The closed conduits could represent air distribution ducts, fluid passages in heat exchangers, or the flow sections inside fluid machinery. The open channels could represent drainage systems, storm sewers, irrigation canals, or natural streams.

The most obvious difference between circular and noncircular sections is the geometry. The diameter D is used as the characteristic dimension of size for pipes and tubes running full under pressure. For both closed and open noncircular sections, the *hydraulic radius R* is used.

11-2   **Hydraulic Radius**

The hydraulic radius of a section is defined as the ratio of the net cross-sectional area of a flow stream to the wetted perimeter of the section. That is,

$$R = \frac{A}{\text{WP}} = \frac{\text{Area}}{\text{Wetted Perimeter}} \tag{11-1}$$

The unit for R should be the foot for the English gravitational unit system.

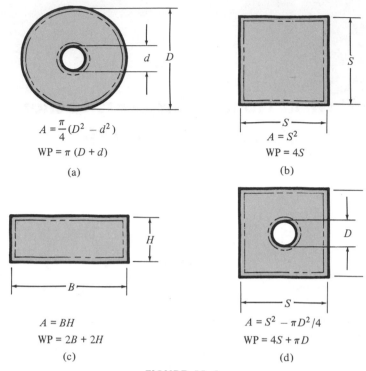

$$A = \frac{\pi}{4}(D^2 - d^2)$$
$$WP = \pi\,(D + d)$$

(a)

$$A = S^2$$
$$WP = 4S$$

(b)

$$A = BH$$
$$WP = 2B + 2H$$

(c)

$$A = S^2 - \pi D^2/4$$
$$WP = 4S + \pi D$$

(d)

FIGURE 11–1

In the calculation of the hydraulic radius, the net cross-sectional area should be evident from the geometry of the section. The wetted perimeter is defined as the sum of the length of the boundaries of the section actually in contact with (that is, wetted by) the fluid. Expressions for the area A and the wetted perimeter WP are given in Fig. 11–1 and 11–2 for those sections illustrated. In each case, the fluid flows in the shaded portion of the section. A dashed line is shown adjacent to the boundaries which make up the wetted perimeter. Notice that the length of the free boundary of an open channel is not included in WP.

Example Problem 11–1. Determine the hydraulic radius of the section shown in Fig. 11–1(d) if the inside dimension of each side of the square is 10 inches and the outside diameter of the tube is 6 inches.

Solution. The net flow area is the difference between the area of the square and the area of the circle.

$$A = S^2 - \pi D^2/4 = (10)^2 - \pi(6)^2/4$$
$$A = 71.7 \text{ in.}^2$$

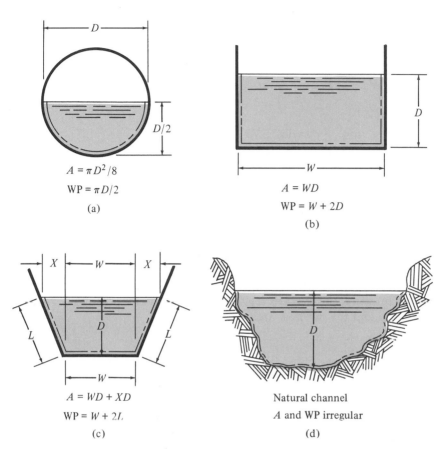

$A = \pi D^2 / 8$

$WP = \pi D / 2$

(a)

$A = WD$

$WP = W + 2D$

(b)

$A = WD + XD$

$WP = W + 2L$

(c)

Natural channel

A and WP irregular

(d)

FIGURE 11-2

The wetted perimeter is the sum of the four sides of the square and the circumference of the circle.

$$WP = 4S + \pi D = 4(10) + \pi(6)$$
$$WP = 58.9 \text{ in.}$$

Then the hydraulic radius R is

$$R = \frac{A}{WP} = \frac{71.7 \text{ in.}^2}{58.9 \text{ in.}} = 1.22 \text{ in.} \left(\frac{1 \text{ ft}}{12 \text{ in.}}\right)$$
$$R = 0.102 \text{ ft}$$

Example Problem 11–2.   Determine the hydraulic radius of the trapezoidal section shown in Fig. 11–2(c) if $W = 4$ ft, $X = 1$ ft, and $D = 2$ ft.

Solution.

$$A = WD + 2(XD/2) = WD + XD$$
$$A = (4)(2) + (1)(2) = 10 \text{ ft}^2$$
$$\text{WP} = W + 2L$$

But,

$$L = \sqrt{X^2 + D^2} = \sqrt{(1)^2 + (2)^2} = 2.24 \text{ ft}$$
$$\text{WP} = 4 + 2(2.24) = 8.48 \text{ ft}$$

Then,

$$R = A/\text{WP} = 10 \text{ ft}^2/8.48 \text{ ft} = 1.18 \text{ ft}$$

11–3   **Closed Noncircular Sections**

When the fluid completely fills the available cross-sectional area and is under pressure, the average velocity of flow is determined by the volume flow rate and the net flow area. That is,

$$v = Q/A \tag{11-2}$$

Also, for steady flow, the volume flow rate is constant throughout the system.

The analysis of flow in closed noncircular sections is similar to the analysis of flow in pipes and tubes if the diameter D is replaced by $4R$. With this one modification, the calculation of Reynolds number and the determination of the energy loss due to friction can be done in the manner discussed in Chapters 6 and 7. The validity of this approach can be demonstrated by calculating the hydraulic radius of a circular pipe.

$$R = \frac{A}{\text{WP}} = \frac{\pi D^2/4}{\pi D} = \frac{D}{4}$$

Then,

$$D = 4R \tag{11-3}$$

The relationships required for calculating friction losses are listed below.
Reynolds number:

$$N_R = \frac{v(4R)\rho}{\mu} = \frac{v(4R)}{\nu} \tag{11-4}$$

Relative roughness $= 4R/\epsilon \tag{11-5}$

Darcy's equation:

$$h_L = f \frac{L}{4R} \frac{v^2}{2g} \tag{11-6}$$

The friction factor f can be determined using the Moody diagram, Fig. 7–2.

Example Problem 11–3. Determine the pressure drop per 100 feet of duct if 5.5 ft³/sec of ethylene glycol at 77°F is flowing through the cross section shown in Fig. 11–1(d). Assume all surfaces are steel having a roughness ϵ of 0.0001 foot. Use the dimensions given in Example Problem 11–1.

Solution. The values for the area and the hydraulic radius were calculated in Example Problem 11–1.

$$A = (71.7 \text{ in.}^2)(1 \text{ ft}^2/144 \text{ in.}^2) = 0.498 \text{ ft}^2$$
$$R = 0.102 \text{ ft}$$

The average velocity of flow is

$$v = \frac{Q}{A} = \frac{5.5 \text{ ft}^3/\text{sec}}{0.498 \text{ ft}^2} = 11.1 \text{ ft/sec}$$

The Reynolds number can now be calculated.

$$N_R = \frac{v(4R)\rho}{\mu} = \frac{(11.1)(4)(0.102)(2.13)}{3.38 \times 10^{-4}}$$
$$N_R = 2.86 \times 10^4$$

The flow is turbulent and Darcy's equation can be used to calculate the energy loss between two points 100 feet apart. In order to determine the friction factor, the relative roughness must first be found.

$$4R/\epsilon = (4)(0.102)/0.0001 = 4080$$

From the Moody diagram, $f = 0.0245$. Then,

$$h_L = f \frac{L}{4R} \frac{v^2}{2g} = \frac{(0.0245)(100)(11.1)^2}{(4)(0.102)(2)(32.2)} \text{ ft}$$
$$h_L = 11.5 \text{ ft}$$

If the duct is horizontal,

$$h_L = \Delta p/\gamma$$
$$\Delta p = \gamma h_L$$

where Δp is the pressure drop caused by the energy loss.

$$\Delta p = \frac{68.47 \text{ lb}}{\text{ft}^3} \cdot 11.5 \text{ ft} \cdot \frac{1 \text{ ft}^2}{144 \text{ in.}^2} = 5.5 \text{ psi}$$

11–4   Classification of Open Channel Flow

Open channel flow occurs when a free surface of the flow stream is exposed to the atmosphere. Since the fluid is not entirely constrained by the channel, the flow is dependent on the slope of the channel as well as its cross-sectional geometry. The fluid motion depends on gravity whereas, in the case of pipe flow, a pressure head could cause the flow.

Throughout this chapter, it will be assumed that the volume flow rate of fluid is constant. The treatment of variable flow rate is quite complicated and beyond the scope of this book. Several references are listed at the end of the chapter if additional information is required. Thus we will be dealing only with steady flow in which the depth of flow is constant at any particular section.

Steady flow can exist as either uniform or varied flow. When uniform flow occurs, the depth is the same at any section along the channel. If there is a change in depth at different positions along the channel, then the flow is called varied.

11–5   Uniform Steady Flow in Open Channels

Figure 11–3 is a schematic illustration of uniform steady flow in an open channel. The distinguishing feature of uniform flow is that the fluid surface is parallel to the slope of the channel bottom. We will use the symbol S_0 to indicate the slope of the channel bottom and S_w for the slope of the water surface. Then for uniform flow, $S_0 = S_w$. Theoretically, uniform flow can exist only if the channel is prismatic, that is, if its sides are parallel to an axis in the direction of flow. Examples of prismatic channels are rectangular, trapezoidal, triangular, and circular sections running partially full. Also, the channel slope S_0 must be constant. If the cross section or slope of the channel is changing, then the flow stream would either be converging or diverging and varied flow would occur.

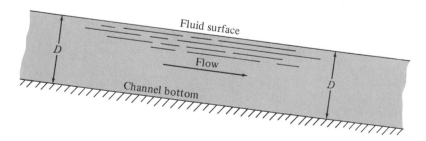

FIGURE 11–3

In uniform flow, the driving force for the flow is provided by the component of the weight of the fluid which acts along the channel as shown in Fig. 11–4. This force is $w \sin \theta$, where w is the weight of a certain element of fluid and θ is the angle of the slope of the channel bottom. If the flow is uniform, it cannot be accelerating. Therefore, there must be an equal opposing force acting along the channel surface. This is a friction force which depends on the roughness of the channel surfaces and on the cross-sectional size and shape.

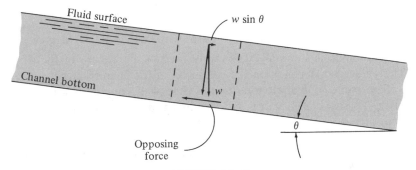

FIGURE 11–4

By equating the expressions for the driving force and the opposing force, an expression for the average velocity of uniform flow can be derived. The most commonly used form of the resulting equation was developed by Robert Manning and is written below.

$$v = \frac{1.49}{n} R^{2/3} S^{1/2} \tag{11-7}$$

In this equation, v is the average velocity of flow, R is the hydraulic radius, S is the channel slope, and n is a resistance factor sometimes called Manning's n.

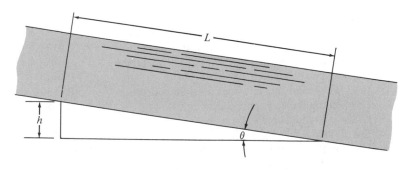

FIGURE 11–5

The slope of a channel can be reported in several ways. It is basically the ratio of the vertical drop to the length of the channel. In Fig. 11–5 this ratio would be h/L. Also, since sin $\theta = h/L$, the angle that the channel bottom makes with the horizontal could be used. Normally the magnitude of the slope for natural streams and drainage structures is very small, a typical value being 0.001. Such a number can also be expressed as a percent, where 0.01 = 1 percent. Then 0.001 = 0.1 percent. In summary, the slope of 0.001 could also be reported as:

1.   The channel falls 1 foot per 1000 feet of channel.
2.   The slope is 0.1 percent.
3.   sin θ = 0.001. Then $\theta = \text{Sin}^{-1}$ (0.001) = 0.057 degree.

Since the angle is so small, it is rarely used as a measure of the slope.

The value of n depends on the condition of the channel surface and is therefore somewhat analogous to the pipe wall roughness ϵ used previously. Typical design values of n are listed in Table 11–1 for materials commonly used for artificial channels and for natural streams. A very extensive discussion of the determination of Manning's n and a more complete table of values are given in the book by V.T. Chow listed with the references at the end of this chapter. The values listed in Table 11–1 are average values which will give good estimates for use in design or for rough analysis of existing channels. Variations from the average should be expected.

TABLE 11–1

Values for Manning's n

Channel description	n
Glass, copper, plastic, or other smooth surfaces	0.010
Smooth unpainted steel, planed wood	0.012
Painted steel or coated cast iron	0.013
Smooth asphalt, common clay drainage tile, trowel finished concrete, glazed brick	0.013
Uncoated cast iron, black wrought iron pipe, vitrified clay sewer	0.014
Brick in cement mortar, float finished concrete	0.015
Formed, unfinished concrete	0.017
Clean excavated earth	0.022
Corrugated metal storm drain	0.024
Earth with light brush	0.050
Earth with heavy brush	0.100

The volume flow rate in the channel can be calculated from the continuity equation which is the same as that used for pipe flow.

$$Q = Av \qquad \qquad (11\text{–}8)$$

In open channel flow work, Q is sometimes called the discharge. Substituting Eq. (11–7) into (11–8) gives an equation which directly relates the discharge to the physical parameters of the channel.

$$Q = \frac{1.49}{n} AR^{2/3}S^{1/2} \qquad \qquad (11\text{–}9)$$

This is the only value of discharge for which uniform flow will occur for the given channel depth and it is called the *normal discharge.*

Another useful form of this equation is shown below.

$$AR^{2/3} = \frac{n\,Q}{1.49\,S^{1/2}} \qquad \qquad (11\text{–}10)$$

The term on the left side of Eq. (11–10) is solely dependent on the geometry of the section. Therefore, for a given discharge, slope, and surface type, the geometrical features of a channel can be determined. Alternatively, for a given size and shape of channel, the equation will allow the calculation of the depth at which the normal discharge Q would occur. This depth is called the *normal depth.*

Typical problems encountered regarding uniform flow are the calculations of the normal discharge, the normal depth, the geometry of the channel section, the slope, or the value of Manning's n. These calculations can be made using Eqs. (11–7) through (11–10).

Example Problem 11–4. Determine the normal discharge for an 8-inch inside diameter common clay drainage tile running half full if it is laid on a slope which drops 1 foot over a run of 1000 feet.

Solution. Equation (11–9) will be used.

$$Q = \frac{1.49}{n} AR^{2/3}S^{1/2}$$

The slope $S = 1/1000 = 0.001$. From Table 11–1 we find $n = 0.013$. Fig. 11–6 shows the cross section of the tile half full.

$$A = \frac{1}{2}\left(\frac{\pi D^2}{4}\right) = \frac{\pi D^2}{8} = 8\pi \text{ in.}^2$$
$$A = 8\pi \text{ in.}^2 \, (1 \text{ ft}^2/144 \text{ in.}^2) = 0.1745 \text{ ft}^2$$
$$WP = \pi D/2 = 4\pi \text{ in.}$$

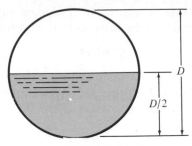

FIGURE 11-6

Then,

$$R = A/WP = 8\pi \text{ in.}^2/4\pi \text{ in.} = 2 \text{ in.}$$

or,

$$R = 0.1667 \text{ ft}$$

Then in Eq. (11-9),

$$Q = \frac{(1.49)(0.1745)(0.1667)^{2/3}(0.001)^{1/2}}{0.013}$$

$$Q = 0.192 \text{ ft}^3/\text{sec}$$

Example Problem 11-5.　Calculate the minimum slope on which the channel shown in Fig. 11-7 must be laid if it is to carry 50 ft³/sec of water with a depth of 2 feet. The sides and bottom of the channel are made of formed, unfinished concrete.

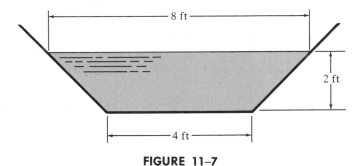

FIGURE 11-7

Solution.　Equation (11-9) can be solved for the slope S.

$$Q = \frac{1.49}{n} AR^{2/3}S^{1/2}$$

$$S = \left[\frac{Qn}{1.49\,AR^{2/3}}\right]^2 \qquad\qquad \textbf{(11-11)}$$

From Table 11–1 we find $n = 0.017$. The values of A and R can be calculated from the geometry of the section.

$$A = (4)(2) + (2)(2)(2)/2 = 12 \text{ ft}^2$$
$$\text{WP} = 4 + 2\sqrt{4 + 4} = 9.66 \text{ ft}$$
$$R = A/\text{WP} = 12/9.66 = 1.24 \text{ ft}$$

Then from Eq. (11–11),

$$S = \left[\frac{(50)(0.017)}{(1.49)(12)(1.24)^{2/3}}\right]^2 = 0.00169$$

Therefore the channel must drop at least 1.69 feet per 1000 feet of length.

Example Problem 11–6. Design a rectangular channel to be made of formed, unfinished concrete to carry 200 ft³/sec of water when laid on a 1.2 percent slope. The normal depth should be one-half the width of the channel bottom.

Solution. Since the geometry of the channel is to be determined, Eq. (11–10) is most convenient.

$$AR^{2/3} = \frac{nQ}{1.49 \ S^{1/2}} = \frac{(0.017)(200)}{1.49 \ (0.012)^{1/2}} = 20.8$$

Figure 11–8 shows the cross section. Since $y = b/2$, only b must be determined. Both A and R can be expressed in terms of b.

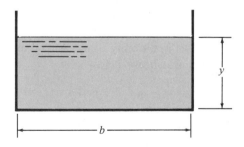

FIGURE 11–8

$$A = by = \frac{b^2}{2}$$
$$\text{WP} = b + 2y = 2b$$
$$R = A/\text{WP} = \frac{b^2}{(2)(2b)} = \frac{b}{4}$$

Then,

$$AR^{2/3} = 20.8$$

$$\frac{b^2}{2}\left(\frac{b}{4}\right)^{2/3} = 20.8$$

$$\frac{b^{8/3}}{5.04} = 20.8$$

$$b = (105)^{3/8} = 5.73 \text{ ft}$$

The width of the channel must be 5.73 ft.

Example Problem 11–7. In the final design of the channel described in Example Problem 11–6, the width was made 6 feet. The maximum expected discharge for the channel is 400 ft³/sec. Determine the normal depth for this discharge.

Solution. Equation (11–10) will be used again.

$$AR^{2/3} = \frac{nQ}{1.49\ S^{1/2}} = \frac{(0.017)(400)}{(1.49)(0.012)^{1/2}} = 41.6$$

Both A and R must be expressed in terms of the dimension y in Fig. 11–8, with $b = 6$ ft.

$$A = 6y$$
$$WP = 6 + 2y$$
$$R = A/WP = 6y/(6 + 2y)$$

Then,

$$41.6 = AR^{2/3} = 6y\left[\frac{6y}{6 + 2y}\right]^{2/3}$$

Algebraic solution for y is not simply done. A trial and error approach can be used. The results are shown below.

y	A	WP	R	$R^{2/3}$	$AR^{2/3}$	
5.0	30.0	16.0	1.875	1.52	45.6	y too high
4.5	27.0	15.0	1.800	1.48	40.0	y too low
4.6	27.6	15.2	1.815	1.49	41.1	y too low
4.65	27.9	15.3	1.825	1.495	41.7	OK

Therefore, the channel depth would be 4.65 feet when the discharge is 400 ft³/sec.

If problems of the type presented in this section are encountered frequently, it is convenient to prepare charts, graphs, or computer programs to facilitate computations. More complete books on open channel flow contain many aids of this type.

11–6   Critical Flow and Specific Energy

The state of flow in an open channel can be predicted by determining whether the velocity of flow is at, above, or below the critical velocity. This can be done by evaluating the dimensionless number N_F called the Froude number.

$$N_F = v/\sqrt{gy_h} \qquad (11\text{--}12)$$

In this equation, v is the average velocity of flow and y_h is the *hydraulic depth*. The value of y_h is calculated by dividing the cross-sectional area of the channel by the width of the free surface, denoted by T.

When $N_F = 1.0$, critical flow occurs. If $N_F < 1.0$, the flow is subcritical and appears smooth and tranquil. If $N_F > 1.0$, the flow is supercritical, has a high velocity, and appears to be quite rapid.

Consideration of energy in open channel flow usually involves a determination of the energy possessed by the fluid at a particular section of interest. The total energy is measured relative to the channel bottom and is composed of potential energy due to the depth of the fluid plus kinetic energy due to its velocity.

Letting E denote the total energy, we get

$$E = y + v^2/2g \qquad (11\text{--}13)$$

where y is the depth and v is the average velocity of flow. As with the energy equation used previously, the terms in Eq. (11–13) have the units of ft-lb per pound of fluid flowing. In open channel flow work, E is usually referred to as the specific energy. For a given discharge Q, the velocity is Q/A. Then

$$E = y + Q^2/2gA^2 \qquad (11\text{--}14)$$

Since the area can be expressed in terms of the depth of flow, Eq. (11–14) relates the specific energy to the depth of flow. A graph of the depth y versus the specific energy E is useful in visualizing the possible regimes of flow in a channel. For a particular channel section and discharge, the specific energy curve appears as shown in Fig. 11–9.

Several features of this curve are important. The 45° line on the graph represents the plot of $E = y$. Then for any point on the curve, the horizontal distance to this line from the y axis represents the potential energy y. The remaining distance to the specific energy curve is the kinetic energy $v^2/2g$. A definite minimum value of E appears and it can be shown that this occurs when the flow is at the critical state, that is, when $N_F = 1$. The depth corresponding to the minimum specific energy is, therefore, called the *critical depth* y_c. For any depth greater than y_c the flow is subcritical. Conversely, for any depth lower than y_c the flow is supercritical. Notice that for any energy level greater than the minimum, there can exist two different depths. In Fig. 11–10, both y_1 below the critical depth y_c, and y_2 above y_c, have the same energy. In the case

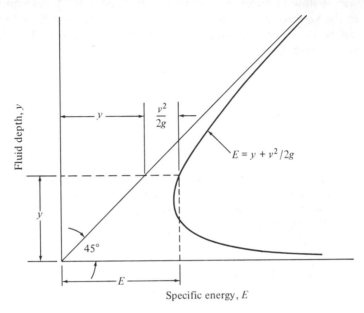

FIGURE 11–9

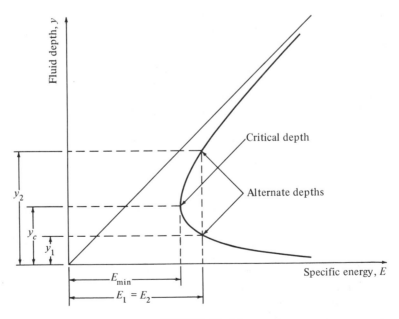

FIGURE 11–10

of y_1, the flow is supercritical and much of the energy is kinetic energy due to the high velocity of flow. At the greater depth y_2, the flow is slower and only a small portion of the energy is kinetic energy. The two depths y_1 and y_2 are called the *alternate depths* for the specific energy E.

11-7	**Hydraulic Jump**

In order to understand the significance of the phenomenon known as hydraulic jump, consider one of its most practical uses illustrated in Fig. 11-11. The water flowing over the spillway normally has a high velocity in the supercritical range when it reaches the bottom of the relatively steep slope at section 1. If this velocity were to be maintained into the natural stream bed beyond the paved spillway structure, the sides and bottom of the stream would be severely eroded. Instead, good design would cause a hydraulic jump to occur as shown, where the depth of flow abruptly changes from y_1 to y_2. Two beneficial effects result from the hydraulic jump. First, the velocity of flow is decreased substantially, decreasing the tendency for the flow to erode the stream bed. Second, much of the excess energy contained in the high velocity flow is dissipated in the jump. Energy dissipation occurs because the flow in the jump is extremely turbulent.

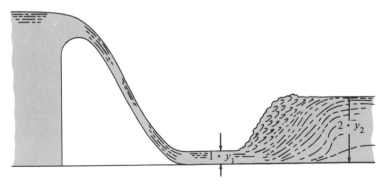

FIGURE 11-11

In order for a hydraulic jump to occur, the flow before the jump must be in the supercritical range. That is, at section 1 in Fig. 11-11, y_1 is less than the critical depth for the channel and the Froude number N_{F_1} is greater than 1.0. The depth at section 2 after the jump can be calculated from

$$y_2 = \tfrac{1}{2}\, y_1(\sqrt{1 + 8N_{F_1}{}^2} - 1) \qquad \textbf{(11-15)}$$

Also, the energy loss in the jump is dependent on the two depths y_2 and y_1.

$$E_1 - E_2 = \Delta E = (y_2 - y_1)^3/4y_1y_2 \qquad (11\text{--}16)$$

Figure 11–12 illustrates what happens in a hydraulic jump using a specific energy curve. The flow enters the jump with an energy E_1 corresponding to a supercritical depth y_1. In the jump, the depth abruptly increases. If no energy were lost, the new depth would be y_2' which is the alternate depth for y_1. However, since there was some energy dissipated, ΔE, the actual new depth y_2 corresponds to the energy level E_2. Still, y_2 is in the subcritical range and tranquil flow would be maintained downstream from the jump. The name given to the actual depth y_2 after the jump is the *sequent depth*.

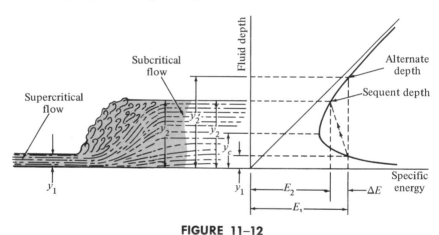

FIGURE 11–12

The example problem below illustrates another practical case in which hydraulic jump might occur.

Example Problem 11–8. As shown in Fig. 11–13, water is being discharged from a reservoir under a sluice gate at the rate of 600 ft³/sec into a horizontal rectangular channel, 10 feet wide, made of unfinished formed concrete. At a point where the depth is 3 feet, a hydraulic jump is observed to occur. Determine the following:

   a) The velocity before the jump
   b) The depth after the jump
   c) The velocity after the jump
   d) The energy dissipated in the jump

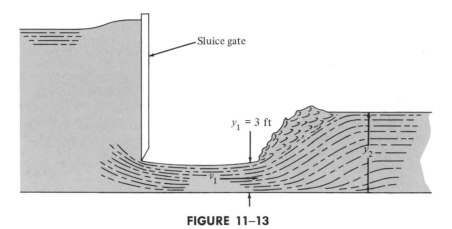

FIGURE 11–13

Solution. a)

$$v_1 = Q/A_1$$
$$A_1 = (10)(3) \text{ ft}^2 = 30 \text{ ft}^2$$
$$v_1 = (600 \text{ ft}^3/\text{sec})/30 \text{ ft}^2 = 20.0 \text{ ft/sec}$$

b) Equation (11–15) can be used to determine the depth after the jump, y_2.

$$y_2 = \tfrac{1}{2} y_1 (\sqrt{1 + 8N_{F_1}^2} - 1)$$
$$N_{F_1} = v_1/\sqrt{gy_h}$$

The hydraulic depth is equal to A/T, where T is the width of the free surface. Then for a rectangular channel $y_h = y$. Then,

$$N_{F_1} = 20/\sqrt{(32.2)(3)} = 2.03$$

The flow is in the supercritical range.

$$y_2 = \tfrac{1}{2}(3)(\sqrt{1 + (8)(2.03)^2} - 1) = 7.25 \text{ ft}$$

c) Because of continuity,

$$v_2 = Q/A_2 = (600 \text{ ft}^3/\text{sec})/(10)(7.25) \text{ ft}^2 = 8.27 \text{ ft/sec}$$

d) From Eq. (11–16),

$$\Delta E = (y_2 - y_1)^3/4y_1 y_2$$
$$\Delta E = \frac{(7.25 - 3.0)^3}{(4)(3.0)(7.25)} \text{ ft} = 0.883 \text{ ft}$$

This means that 0.883 ft-lb of energy is dissipated from each pound of water as it flows through the jump.

References

Albertson, M. L., J. R. Barton, and D. B. Simons. *Fluid Mechanics for Engineers.* Englewood Cliffs, N.J.: Prentice-Hall, Inc., 1960.

Binder, R. C. *Fluid Mechanics.* Englewood Cliffs, N.J.: Prentice-Hall, Inc., Fourth Edition, 1962.

Chow, V. T. *Open Channel Hydraulics.* New York: McGraw-Hill Book Co., 1959.

Henderson, F. M. *Open Channel Flow.* New York: The Macmillan Co., 1966.

PRACTICE PROBLEMS

11–1 Water at 120°F flows in the annular space between a 4-inch Type K copper tube and a 1½-inch Type K copper tube. The tubes are arranged in the manner shown in Fig. 11–1(a). Calculate the pressure drop per 100 feet if the flow rate is 600 gal/min.

11–2 Air with a specific weight of 0.08 lb/ft³ and a dynamic viscosity of 4.0×10^{-7} lb-sec/ft² flows through the shaded portion of the duct shown in Fig. 11–14 at the rate of 100 ft³/min. Calculate the Reynolds number of the flow.

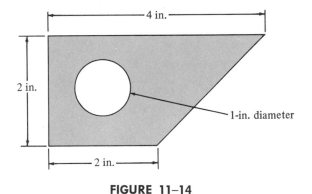

FIGURE 11–14

11–3 Carbon dioxide having a specific weight of 0.114 lb/ft³ and a dynamic viscosity of 3.34×10^{-7} lb-sec/ft² flows in the shaded portion of the duct shown in Fig. 11–15. If the volume flow rate is 200 ft³/min calculate the Reynolds number of the flow.

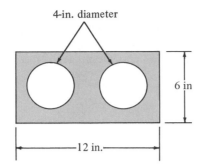

4-in. diameter

6 in

12 in.

FIGURE 11–15

11–4 Glycerine (sg = 1.26) at 100°F flows in the shaded portion of the duct shown in Fig. 11–16. Calculate the Reynolds number of the flow if the flow rate is 3.50 ft³/sec.

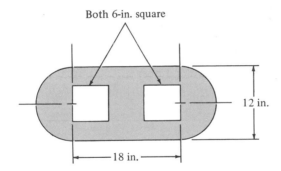

Both 6-in. square

12 in.

18 in.

FIGURE 11–16

11–5 A rectangular heating duct, 12 in. × 18 in., carries air at a velocity of 6 ft/sec. Consider the air to be incompressible with a kinematic viscosity of 2.0×10^{-4} ft²/sec. The duct is galvanized iron having a wall roughness of 0.0005 foot. Calculate the energy loss due to friction per 100 feet of duct.

11–6 Water at 90°F flows in the space between a 6-inch diameter pipe and a 10-inch square duct. The wall roughness is 0.004 foot for all surfaces. Find the maximum volume flow rate if the maximum allowable pressure drop is 5 psi per 100 feet.

11–7 Water at 60°F is flowing through a horizontal shell and tube heat exchanger at the rate of 0.50 ft³/sec. The cross section of the heat exchanger is shown in Fig. 11–17 and the flow is in the shaded area. The inside diameter of the shell is 2 inches. The outside diameter of each tube is $\frac{1}{2}$ inch. The material is steel having an average roughness of 0.0001 foot. Calculate the pressure drop in the heat exchanger if it is 5 feet long.

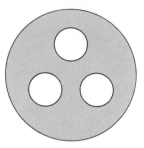

FIGURE 11–17

11–8 Figure 11–18 shows a heat exchanger used to cool a bank of electronic devices. Ethylene glycol at 77°F is flowing in the shaded area. The total length of the heat exchanger is 3.5 feet. For this system, calculate the volume flow rate required to produce a Reynolds number of 1500. At this flow rate calculate the pressure drop across the heat exchanger due to friction.

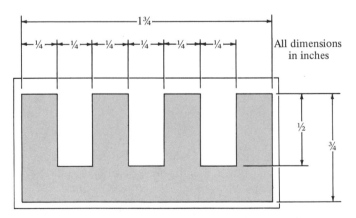

FIGURE 11–18

11–9 A heat exchanger is made up of a rectangular passage with five tubes arranged perpendicular to the direction of flow as shown in Fig. 11–19. Each tube is $\frac{1}{2}$ inch in diameter. Air at atmospheric pressure and 70°F is flowing at 20 ft³/min. Cal-

culate the hydraulic radius and the Reynolds number at the
section of minimum area. (See Table 14–2 for the properties of
air.)

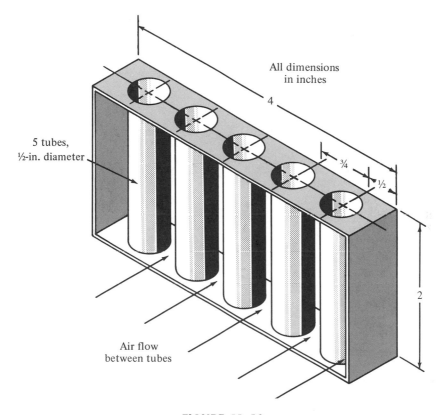

All dimensions
in inches

5 tubes,
½-in. diameter

4

¾

½

2

Air flow
between tubes

FIGURE 11–19

11–10   Water is flowing in a formed unfinished concrete rectangular
channel, 10 feet wide. For a depth of 6 feet calculate the normal
discharge and the Froude number of the flow. The channel
slope is 0.1 percent.

11–11   Determine the normal discharge for an aluminum rain spout
having the shape shown in Fig. 11–20 when running at the
depth of 3 inches. Use $n = 0.013$. The spout falls 4 inches
over a length of 60 feet.

11–12   A culvert under a highway is 6 feet in diameter and is made of
corrugated metal. It drops 1 foot over a length of 500 feet. Calculate the normal discharge when the culvert runs half full.

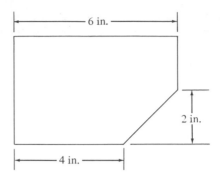

FIGURE 11–20

11–13 A wooden flume is being built to temporarily carry 1250 gal/ min of water until a permanent drain can be installed. The flume is rectangular having an 8.25 inch bottom width and a maximum depth of 10 inches. Calculate the required slope to handle the expected discharge.

11–14 A storm drainage channel in a city where heavy sudden rains occur has the shape shown in Fig. 11–21. It is made of unfinished concrete and has a slope of 0.5 percent. During normal times, the water remains in the small rectangular section. The upper section allows large volumes to be carried by the channel. Determine the normal discharge for depths of 1.5 feet and 8 feet.

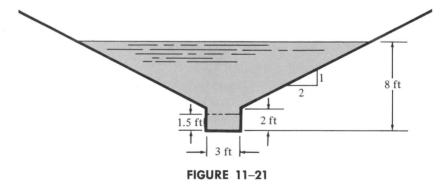

FIGURE 11–21

11–15 Figure 11–22 represents the approximate shape of a natural stream channel with levees built on either side. The channel is earth with grass cover. Use $n = 0.04$. If the average slope is 0.00015, determine the normal discharge for depths of 3 feet and 6 feet.

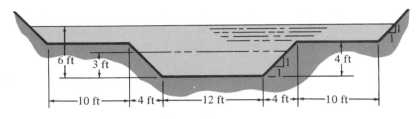

FIGURE 11–22

11–16 Calculate the depth of flow of water in a rectangular channel, 10 feet wide, made of brick in cement mortar for a discharge of 150 ft³/sec. The slope is 0.1 percent.

11–17 Calculate the depth of flow in a trapezoidal channel having a bottom width of 10 feet and whose walls slope 45° with the horizontal. The channel is made of unfinished concrete and is laid on a 0.1 percent slope. The discharge is 500 ft³/sec.

11–18 A rectangular channel is to carry 60 ft³/sec of water from a water cooled refrigeration condenser to a cooling pond. The available slope is 6 inches over a distance of 500 feet. The maximum depth of flow is 1 foot. Determine the width of the channel if its surface is trowel finished concrete.

11–19 The channel shown in Fig. 11–23 has a surface of float finished concrete and is laid on a slope which falls 1 foot in 1000 feet of length. Calculate the normal discharge and the Froude number for a depth of 5 feet. For that discharge calculate the critical depth.

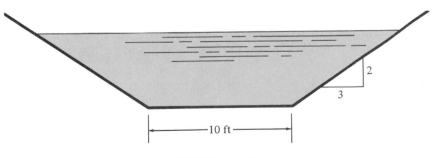

FIGURE 11–23

11–20 A square storage room is equipped with automatic sprinklers for fire protection which spray 1000 gal/min of water. The floor is designed to drain this flow evenly to troughs near each outside wall shaped as shown in Fig. 11–24. Each trough is laid on a 1 percent slope and is formed of unfinished concrete. Determine the minimum depth h.

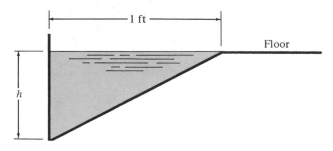

FIGURE 11–24

11–21 The flow from two troughs as described in problem 11–20 passes into a sump from which a round common clay drainage tile carries it to a storm sewer. Determine the required size of tile to carry the flow (500 gal/min) when running half full and the slope is 0.1 percent.

11–22 A rectangular channel 6 feet wide carries 200 ft³/sec and is made of unfinished concrete. Perform the following operations.

(a) Calculate the critical depth.

(b) Calculate the minimum specific energy.

(c) Plot the specific energy curve.

(d) Determine the specific energy when $y = 2.0$ feet and the alternate depth for this energy.

(e) Determine the velocity of flow and the Froude numbers for each depth in (d) above.

(f) Calculate the required slopes of the channel if the depths from (d) above are to be normal depths for 200 ft³/sec flow.

12

Flow Measurement

The Need for Flow Measurement

The three primary reasons for using flow measurement systems are accounting, performance evaluation, and process control. Whenever the custody of a fluid is transferred there is a need to account for the quantity involved. In commerce, examples of custody transfer are numerous. Gasoline flow is measured as it is pumped into the fuel tank of a car. Natural gas for cooking and heating is metered, along with water use. In industry, both the raw material input and finished product output must be continuously monitored for internal accounting purposes. Examples of performance evaluation are the measurement of fuel flow into an engine, air flow in a heating system, blood flow during surgery, or water flow through a heat exchanger. In process control, the success of a continuous operation is very dependent on flow measurement and control. For example, in paper making the flow of the pulp slurry into the machine and the flow of steam to the drying rolls must be monitored and controlled to ensure a uniform product. In the manufacture of urethane foam insulation, the density and thus the insulating quality of the foam is dependent on the precise metering of two primary constituents.

Many devices are available for measuring flow. Some measure volume flow rate directly, while others measure an average velocity of flow which can then be converted to volume flow rate by using the continuity equation, $Q = Av$. Also, some provide direct primary measurements while others require calibration or the application of a discharge coefficient to the observed output of the device. The form of the flow meter output also varies considerably from one type to another. The indication can be a pressure, a liquid level, a mechanical counter, the position of an indicator in the fluid stream, a continuous electrical signal, or a series of electrical pulses. The choice of the basic type of fluid meter and its indication system depends on several factors, some of which are discussed below.

Range. Commercially available meters can measure flows from a few cubic inches per hour for precise laboratory experiments to several thousand cubic feet per second for irrigation water or municipal water and sewage systems. Then for a particular meter installation, the general order of magnitude of the flow rate must be known as well as the range of the expected variations.

Accuracy Required. Virtually any flow measuring device, properly installed and operated, can produce an accuracy within 5 percent of the actual flow. Most commercially made meters are capable of 2 percent accuracy, and several claim better than $\frac{1}{2}$ percent. Cost usually becomes an important factor when great accuracy is desired.

Pressure Loss. Because the construction details of the various meters are quite different, they produce differing amounts of energy loss or pressure loss as the fluid flows through them. Except for a few types, fluid meters accomplish the measurement by placing a restriction or a mechanical device in the flow stream, thus causing the energy loss.

Type of Indication. Factors to consider regarding the type of flow indication include whether remote sensing or recording is required, whether automatic control is to be actuated by the output, whether an operator needs to monitor the output, and whether severe environmental conditions exist.

Type of Fluid. The performance of some fluid meters is affected by the properties and condition of the fluid. A basic consideration is whether the fluid is a liquid or a gas. Other factors which may be important are viscosity, temperature, corrosiveness, electrical conductivity, optical clarity, lubricating properties, and homogeneity.

In most cases the physical size of the meter, cost, the system pressure, and the operator's skill should also be considered.

Calibration is required for some types of flow meters. Some manufacturers provide a calibration in the form of a graph or chart of actual flow versus indicator reading. Some are equipped for direct reading having scales calibrated in the desired units of flow. In the case of the more fundamental types of meters, such as the variable head types, standard geometrical forms and dimensions have been determined for which empirical data are available relating flow to an easily measured variable such as a pressure difference or fluid level. References at the end of this chapter give many of these calibration factors.

If calibration is required by the user of the device he may use another precision meter as a standard against which the reading of the test device can be compared. Alternatively, primary calibration can be performed by adjusting the flow to a constant rate through the meter and then collecting the output during a fixed time interval. The fluid thus collected can either be weighed for a weight per unit time calibration, or its volume can be measured for a volume flow rate calibration.

12–2	**Variable Head Meters**

The basic principle on which variable head meters are based is that when a fluid stream is restricted, its pressure decreases by an amount which is dependent on the rate of flow through the restriction. Therefore, the pressure difference between points before and after the restriction can be used to indicate flow rate. The most common types of variable head meters are the venturi tube, the flow nozzle, and the orifice. The derivation of the relationship between the pressure difference and the volume flow rate is the same regardless of which type of device is used. The venturi tube will be used as an example.

Venturi Tube Figure 12–1 shows the basic appearance of a venturi tube. The flow from the main pipe at section 1 is caused to accelerate through a narrow section called the throat where the fluid pressure is decreased. The flow then expands through the diverging portion to the same diameter as the main pipe. Pressure taps are located in the pipe wall at section 1 and in the wall of the throat, here called section 2. These pressure taps are attached to the two sides of a differential manometer so that the deflection h is an indication of the pressure difference $p_1 - p_2$. Of course, other types of differential pressure gages could be used.

The energy equation and the continuity equation can be used to derive the relationship from which the flow rate can be calculated. Using sections 1 and 2 in Fig. 12–1 as the reference points, the two equations are:

$$\frac{p_1}{\gamma} + z_1 + \frac{v_1^2}{2g} - h_L = \frac{p_2}{\gamma} + z_2 + \frac{v_2^2}{2g} \qquad (12\text{–}1)$$

$$Q = A_1 v_1 = A_2 v_2 \qquad (12\text{–}2)$$

These equations are valid only for incompressible fluids, that is, liquids. For the flow of gases, special consideration of the variation of the

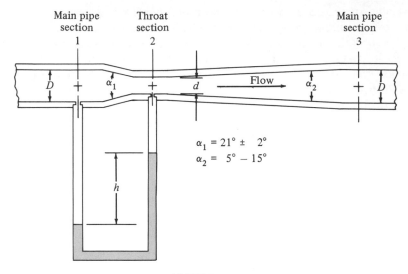

FIGURE 12–1

Venturi Tube

specific weight γ with pressure must be given. The algebraic reduction of Eqs. (12–1) and (12–2) proceeds as follows.

$$\frac{v_2^2 - v_1^2}{2g} = \frac{p_1 - p_2}{\gamma} + (z_1 - z_2) - h_L$$

$$v_2^2 - v_1^2 = 2g[(p_1 - p_2)/\gamma + (z_1 - z_2) - h_L]$$

But $v_1^2 = v_2^2(A_2/A_1)^2$. Then,

$$v_2^2[1 - (A_2/A_1)^2] = 2g[(p_1 - p_2)/\gamma + (z_1 - z_2) - h_L]$$

$$v_2 = \sqrt{\frac{2g[(p_1 - p_2)/\gamma + (z_1 - z_2) - h_L]}{1 - (A_2/A_1)^2}} \quad \textbf{(12–3)}$$

Two simplifications can be made at this time. First, the elevation difference $(z_1 - z_2)$ is very small, even if the meter is installed vertically. Therefore, this term is neglected. Second, the term h_L is the energy loss from the fluid as it flows from section 1 to section 2. The value of h_L must be determined experimentally. But it is more convenient to modify Eq. (12–3) by dropping h_L and introducing a discharge coefficient C.

$$v_2 = C\sqrt{\frac{2g(p_1 - p_2)/\gamma}{1 - (A_2/A_1)^2}} \quad \textbf{(12–4)}$$

This equation can be used to calculate the velocity of flow in the throat of the meter. However, normally the volume flow rate is desired. Since $Q = A_2 v_2$,

$$Q = CA_2 \sqrt{\frac{2g(p_1 - p_2)/\gamma}{1 - (A_2/A_1)^2}} \tag{12-5}$$

The value of the coefficient C is dependent on the Reynolds number of the flow and on the actual geometry of the meter. For a well designed venturi tube, C is approximately 0.984 for Reynolds numbers of 2×10^5 and larger in the main pipe. Figure 12–2 shows a typical curve of C versus Reynolds number.

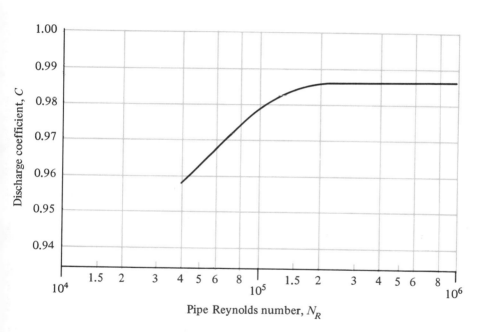

FIGURE 12–2

Venturi Tube Discharge Coefficient

Source: ASME
  Fluid Meters: Their Theory and Application

Equation (12–5) is used for the flow nozzle and the orifice as well as the venturi tube. These devices are described below.

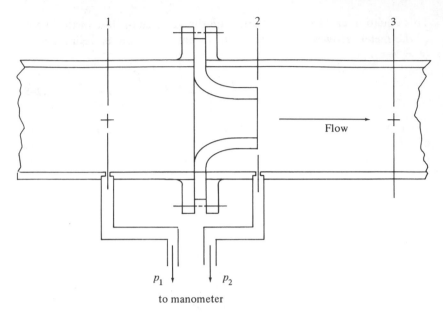

FIGURE 12–3

Flow Nozzle

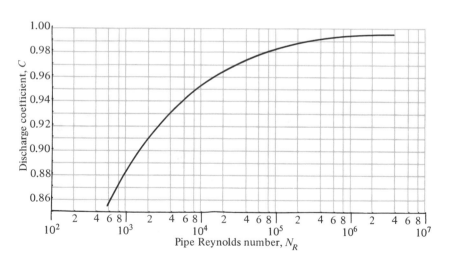

FIGURE 12–4

Flow Nozzle Discharge Coefficient

Source: ASME
Fluid Meters: Their Theory and Application

Flow Nozzle. The flow nozzle is a gradual contraction of the flow stream followed by a short straight cylindrical section as illustrated in Fig. 12–3. Several standard geometries have been presented and adopted by organizations such as the American Society of Mechanical Engineers and the International Organization for Standardization. Because of the smooth gradual contraction, there is very little energy loss between points 1 and 2 for a flow nozzle. The value of *C* is thus very close to one. A typical curve of *C* versus Reynolds number is shown in Fig. 12–4.

Orifice. A flat plate with an accurately machined, sharp-edged hole is referred to as an orifice. Normally the orifice is placed concentrically inside the pipe between flanges as shown in Fig. 12–5. As the flow stream encounters the plate, it converges and the sharp edge causes a vena contracta to form some distance downstream from the plate. The actual value of the discharge coefficient *C* depends on the location of the pressure taps with three possible locations listed in Table 12–1.

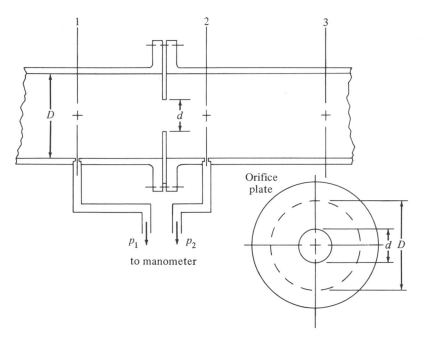

FIGURE 12–5

Square-Edged Orifice

The value of *C* also is affected by small variations in the geometry of the edge of the orifice. Typical curves for sharp-edged orifices are

TABLE 12–1

	Inlet pressure tap p_1	Outlet pressure tap p_2
1	One pipe diameter up-stream from plate	One-half pipe diameter downstream from plate
2	One pipe diameter up-stream from plate	At vena contracta
3	In flange, 1 in. upstream from plate	In flange, 1 in. downstream from plate

shown in Fig. 12–6, where D is the pipe diameter and d is the orifice diameter. The value of C is much lower than that for the venturi tube or the flow nozzle since the fluid is forced to make a sudden contraction. Also, since measurements are based on the orifice diameter, the de-crease in the diameter of the flow stream at the vena contracta tends to reduce the value of C.

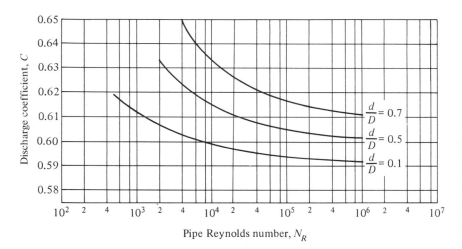

FIGURE 12–6

Orifice Discharge Coefficient

Source: ASME
Fluid Meters: Their Theory and Application

Pressure Recovery. In each of the three types of variable head meters described above, the flow stream expands back to the main pipe diameter after passing the restriction. This is indicated as section 3 in Figs. 12–1, 12–3, and 12–5. Then the difference between the pressures p_1 and p_3 is due to the meter. The difference can be evaluated by considering the energy equation.

$$\frac{p_1}{\gamma} + z_1 + \frac{v_1^2}{2g} - h_L = \frac{p_3}{\gamma} + z_3 + \frac{v_3^2}{2g}$$

Because the pipe sizes are the same at both sections, $v_1 = v_3$. We may also assume $z_1 = z_3$. Then,

$$p_1 - p_3 = \gamma h_L \qquad (12\text{--}6)$$

The pressure drop is proportional to the energy loss. The careful streamlining of the venturi tube and the long gradual expansion after the throat cause very little excess turbulence in the flow stream. Therefore, energy loss is low and pressure recovery is high. The lack of a gradual expansion causes the nozzle to have a lower pressure recovery while that for an orifice is still lower.

12–3   Variable Area Meters

The rotameter is a common type of variable area meter. Figure 12–7 shows a typical geometry. The fluid flows upward through a clear tube which has an accurate taper on the inside. A float is suspended in the flowing fluid at a position proportional to the flow rate. The upward forces due to the fluid dynamic drag on the float and buoyancy just balance the weight of the float. A different flow rate causes the float to move to a new position, changing the clearance area between the float and the tube until equilibrium is achieved again. The position of the float is measured against a calibrated scale, graduated in convenient units of volume flow rate or weight flow rate.

12–4   Turbine Flow Meter

Figure 12–8 shows a turbine flow meter in which the fluid causes the turbine rotor to rotate at a speed dependent on the flow rate. As each blade of the rotor passes the magnetic coil, a voltage pulse is generated which can be input to a frequency meter, electronic counter, or other similar device whose readings can be converted to flow rate. Flow rates from as low as 0.005 gallons per minute to several thousand gallons per minute can be measured with turbine flow meters of various sizes.

FIGURE 12–7

Rotameter

Source: Fischer & Porter Co.

FIGURE 12–8

Turbine Flow Meter

Source: Flow Technology, Inc.

12–5   Magnetic Flow Meter

Totally unobstructed flow is one of the advantages of magnetic flow meters like that shown in Fig. 12–9. The fluid must be slightly conducting since the meter operates on the principle that when a moving conductor cuts across a magnetic field, a voltage is induced. The primary components of the magnetic flow meter include a tube lined with a nonconducting material, two electromagnetic coils, and two electrodes mounted 180° apart in the tube wall. The electrodes detect the voltage generated in the fluid. Since the generated voltage is directly proportional to the fluid velocity, a greater flow rate generates a greater voltage. An important feature of this type of meter is that its output is completely independent of temperature, viscosity, specific gravity, or turbulence. Tube sizes from 0.2 inch to several feet in diameter are available.

FIGURE 12–9

Magnetic Flow Meter

Source: The Foxboro Co.

12–6   Velocity Probes

Several devices are available which measure velocity of flow at a specific location rather than an average velocity. These are referred to as velocity probes. Some of the more common types will be described in this section.

Pitot Tube.   When a moving fluid is caused to stop because it encounters a stationary object, a pressure is created which is greater than the pressure of the fluid stream. The magnitude of this increased pressure is related to the velocity of the moving fluid. The pitot tube uses this principle to indicate velocity as illustrated in Fig. 12–10. The pitot tube is a hollow tube positioned so that the open end points directly

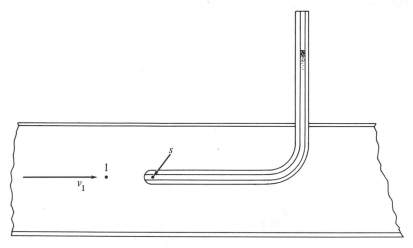

FIGURE 12–10

into the fluid stream. The pressure at the tip causes a column of fluid to be supported. The fluid at or just inside the tip is then stationary or stagnant and the point is referred to as the stagnation point. The energy equation can be used to relate the pressure at the stagnation point with the fluid velocity. If point 1 is taken in the undisturbed stream ahead of the tube and point s is taken at the stagnation point, then,

$$\frac{p_1}{\gamma} + z_1 + \frac{v_1^2}{2g} - h_L = \frac{p_s}{\gamma} + z_s + \frac{v_s^2}{2g} \qquad (12\text{–}7)$$

Observe that $v_s = 0$, $z_1 = z_2$ or very nearly so, and $h_L = 0$ or very nearly so. Then,

$$\frac{p_1}{\gamma} + \frac{v_1^2}{2g} = \frac{p_s}{\gamma} \qquad (12\text{–}8)$$

The names given to the terms in Eq. (12–8) are as follows:

$$p_1 = \text{Static pressure in the main fluid stream}$$
$$p_1/\gamma = \text{Static pressure head}$$
$$p_s = \text{Stagnation pressure or total pressure}$$
$$p_s/\gamma = \text{Total pressure head}$$
$$v_1^2/2g = \text{Velocity pressure head}$$

Then the total pressure head is equal to the sum of the static pressure head and the velocity pressure head. Solving Eq. (12–8) for the velocity gives

$$v_1 = \sqrt{2g(p_s - p_1)/\gamma} \qquad (12\text{–}9)$$

Notice that only the difference between p_s and p_1 is required to calculate the velocity. For this reason, most pitot tubes are made as shown in

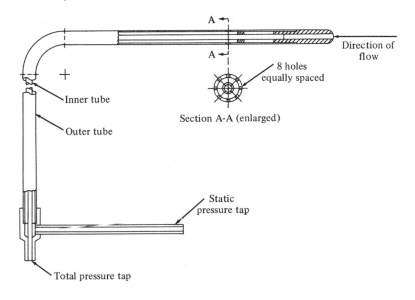

Section A-A (enlarged)

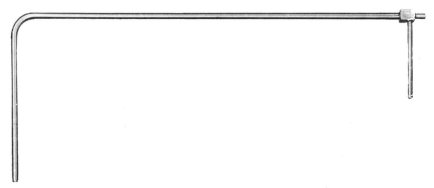

FIGURE 12–11

Pitot Tube

Source: F. W. Dwyer Co.

Fig. 12–11, providing for the measurement of both pressures with the same device. If a differential manometer is used as shown in Fig. 12–12, the manometer deflection h can be related directly to the velocity. The

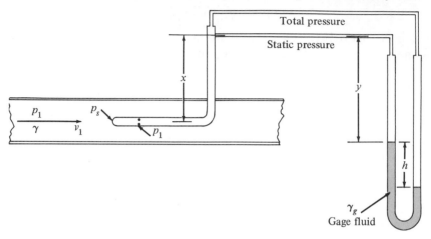

FIGURE 12–12

equation describing the difference between p_s and p_1 can be written by starting at the static pressure taps in the side of the tube, proceeding through the manometer, and ending at the open tip of the tube at point s.

$$p_1 - \gamma x + \gamma y + \gamma_g h - \gamma h - \gamma y + \gamma x = p_s$$

The terms involving the unknown distances x and y drop out. Then, solving for the pressure difference,

$$p_s - p_1 = \gamma_g h - \gamma h = h(\gamma_g - \gamma) \qquad \text{(12–10)}$$

This can be substituted into Eq. (12–9), giving,

$$v_1 = \sqrt{2gh(\gamma_g - \gamma)/\gamma} \qquad \text{(12–11)}$$

The velocity calculated by either Eq. (12–9) or (12–11) is the local velocity at the particular location of the tip of the tube. It was stated in Chapter 6 that the velocity of flow varies from point to point across a pipe. Therefore, if the average velocity of flow is desired, a traverse of the pipe should be made with the tip of the tube placed at the ten points indicated in Fig. 12–13. The dashed circles define concentric annular rings which have equal areas. The velocity at each point can be calculated using Eq. (12–11). Then the average velocity of flow is the average of these ten values. The volume flow rate can be found from $Q = Av$, using the average velocity.

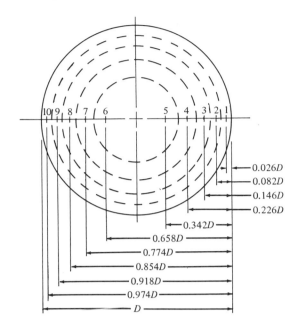

FIGURE 12-13

Example Problem 12-1. For the apparatus shown in Fig. 12–12, the fluid in the pipe is water at 140°F and the manometer fluid is mercury having a specific gravity of 13.60. If the manometer deflection h is 10.4 inches, calculate the velocity of the water.

Solution. Equation (12–11) will be used.

$$v_1 = \sqrt{2gh(\gamma_g - \gamma)/\gamma}$$
$$\gamma = 61.4 \text{ lb/ft}^3 \qquad \text{(water at 140°F)}$$
$$\gamma_g = (13.60)(62.4 \text{ lb/ft}^3) = 848 \text{ lb/ft}^3 \qquad \text{(Mercury)}$$
$$h = 10.4 \text{ in. (1 ft/12 in.)} = 0.867 \text{ ft}$$

Since all terms are in the lb-ft-sec unit system, the velocity is in ft/sec.

$$v_1 = \sqrt{\frac{(2)(32.2)(0.867)(848 - 61.4)}{61.4}}$$
$$v_1 = 26.8 \text{ ft/sec}$$

Cup Anemometer. Air velocity is often measured with a cup anemometer such as that shown in Fig. 12–14. The moving air strikes the open cups causing rotation of the shaft on which they are mounted. The rotational speed of the shaft is proportional to the air velocity which is indicated on a meter or transmitted electrically.

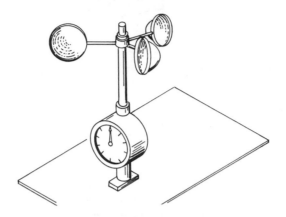

FIGURE 12–14

Rotating Cup Anemometer

Hot Wire Anemometer.   This type of velocity probe employs a very thin wire, about 0.0005 inch diameter, through which an electric current is passed. The wire is suspended on two supports as shown in Fig. 12–15 and inserted into the fluid stream. The wire tends to heat because of the current flowing in it, but it is cooled by convection heat transfer to the moving fluid stream. The amount of cooling depends on the velocity of the fluid. In one type of hot wire anemometer, a constant current is applied to the wire. A variation in the flow velocity causes a change in the wire temperature and therefore its resistance changes. The electronic measurement of the resistance change can be related to flow velocity. Another type senses a change in the resistance of the wire but then varies the current flow to maintain a set wire temperature regardless of fluid velocity. The magnitude of the current flow is then related to fluid velocity.

Hot wire

FIGURE 12–15

Hot Wire Anemometer Tip

12-7   Open Channel Flow Measurement

Two widely used devices for open channel flow measurement are weirs and flumes. Each causes the area of the stream to change which in turn changes the level of the fluid surface. The resulting level of the surface relative to some feature of the device is related to the quantity of flow. Large volume flow rates of liquids can be measured with weirs and flumes.

Weirs. A weir is a barrier or dam placed in the channel so that the fluid backs up behind it and then falls through a notch cut into the face of the weir. Two common notch geometries are the rectangular and the triangular. Figure 12–16 shows a side view of a weir in operation. Front views showing different notch geometries are in Fig. 12–17. The discharge over the weir is dependent on the dimensions of the notch and on the head H of the fluid. Figure 12–16 shows that the fluid surface is somewhat curved as it passes over the crest of the weir. In order to

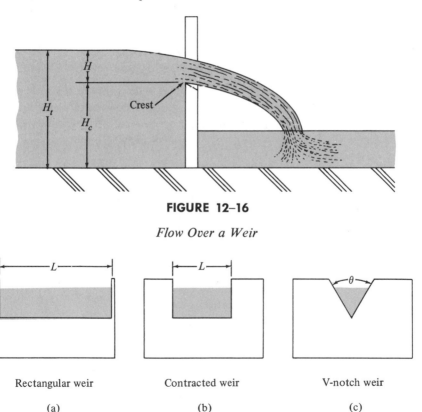

FIGURE 12–16

Flow Over a Weir

Rectangular weir	Contracted weir	V-notch weir
(a)	(b)	(c)

FIGURE 12–17

ensure consistent measurements, the value of H should be the difference between the height to the crest H_c and the total height to the liquid surface H_t, with H_t measured upstream from the weir plate where the surface profile is undisturbed. Normally this upstream distance is approximately six times the maximum expected head H.

The theoretical equation for discharge over a rectangular notched weir is

$$Q = \tfrac{2}{3}L\sqrt{2gH^3} \qquad\qquad (12\text{--}12)$$

where L is the length of the crest between the sides of the notch and H is the head above the crest. If L and H are measured in feet then Q is in ft³/sec.

The actual discharge is different from the theoretical for a variety of reasons and therefore, more accurate and easier to use relationships have been developed. Most formulas take the form,

$$Q = C L H^{3/2} \qquad\qquad (12\text{--}13)$$

where C is a discharge coefficient, L is the effective length of the crest and H is the head above the weir crest. For the full width rectangular weir in Fig. 12–17(a), the following equation can be used:

$$Q = (3.27 + 0.40\, H/H_c)L\, H^{3/2} \qquad\qquad (12\text{--}14)$$

The two end contractions in the contracted weir in Fig. 12–17(b) cause the stream to be curved in from the sides, decreasing the effective length of the crest. The discharge for this type of weir can be calculated from

$$Q = (3.27 + 0.40H/H_c)(L - 0.2H)\, H^{3/2} \qquad\qquad (12\text{--}15)$$

The triangular weir is used primarily for low flow rates since the V-notch produces a larger head H than would be obtained with a rectangular notch. The angle of the V-notch is a factor in the discharge equation. Angles from $35°$ to $120°$ are satisfactory, but $60°$ and $90°$ are quite commonly used. The theoretical equation for a triangular weir is

$$Q = \tfrac{8}{15}C\sqrt{2g}\, \tan (\theta/2)H^{5/2} \qquad\qquad (12\text{--}16)$$

where θ is the total included angle between the sides of the notch. An additional reduction of this equation gives

$$Q = 4.28\, C \tan (\theta/2)\, H^{5/2} \qquad\qquad (12\text{--}17)$$

The value of C is somewhat dependent on the head H, but a nominal value is 0.58. Using this and the common values of $60°$ and $90°$ for θ gives

$$Q = 1.43\, H^{5/2} \text{ (60° notch)} \qquad\qquad (12\text{--}18)$$
$$Q = 2.48\, H^{5/2} \text{ (90° notch)} \qquad\qquad (12\text{--}19)$$

Flumes. Critical flow flumes are contractions in the stream which cause the flow to achieve its critical depth within the structure. There is a definite relationship between depth and discharge when critical flow exists. A widely used type of critical flow flume is the Parshall flume, the geometry of which is shown in Fig. 12–18. The discharge is dependent on the width of the throat section L and the head H, where H is measured at a specific location along the converging section of the flume. The discharge equations in Table 12–2 were developed empirically.

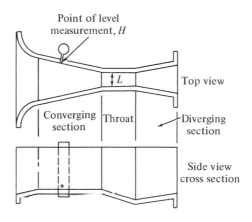

FIGURE 12–18

Parshall Flume

TABLE 12–2

Discharge Equations for Parshall Flumes

Throat width L	Flow range (ft³/sec)		Equation H and L in ft Q in ft³/sec
	Min.	Max.	
3 in.	0.03	1.9	$Q = 0.992\ H^{1.547}$
6 in.	0.05	3.9	$Q = 2.06\ H^{1.58}$
9 in.	0.09	8.9	$Q = 3.07\ H^{1.53}$
1 ft	0.11	16.1	
2 ft	0.42	33.1	
4 ft	1.3	67.9	$Q = 4L\ H^{1.522L^{0.26}}$
6 ft	2.6	103.5	
8 ft	3.5	139.5	
10 ft	6	200	
20 ft	10	1000	
30 ft	15	1500	$Q = (3.6875L + 2.5)\ H^{1.6}$
40 ft	20	2000	
50 ft	25	3000	

References

ASME Research Committee on Fluid Meters. *Fluid Meters: Their Theory and Application.* New York: The American Society of Mechanical Engineers, (5th ed.), 1959.

ASME Research Committee on Fluid Meters. *Flowmeter Computation Handbook.* New York: The American Society of Mechanical Engineers, 1961.

Beckwith, T. G., and N. L. Buck. *Mechanical Measurements.* Reading, Mass.: Addison-Wesley Publishing Co., Inc., 1961.

Chow, V. T. *Open-Channel Hydraulics.* New York: McGraw-Hill Book Co., 1959.

Considine, D. M., Editor-in-Chief. *Handbook of Applied Instrumentation.* New York: McGraw-Hill Book Co., 1964.

Jones, E.B. *Instrument Technology.* London: Butterworth & Co., Ltd., (2nd ed.), 1965.

Spink, L. K. *Principles and Practice of Flow Meter Engineering.* Foxboro, Mass.: The Foxboro Company, (9th ed.), 1967.

Streeter, V. L., Editor-in-Chief. *Handbook of Fluid Dynamics.* New York: McGraw-Hill Book Co., 1961..

PRACTICE PROBLEMS

12-1   A venturi meter similar to the one in Fig. 12-1 has a pipe diameter of 4 inches and a throat diameter of 2 inches. While carrying water at 180°F a pressure difference of 8.0 psi is observed between sections 1 and 2. Calculate the volume flow rate of water.

12-2   Air having a specific weight of 0.081 lb/ft^3 and a kinematic viscosity of 1.42×10^{-4} ft^2/sec is flowing through a flow nozzle similar to that shown in Fig. 12-3. A manometer using water as the gage fluid reads 3.2 inches of deflection. Calculate the volume flow rate if the nozzle diameter is 2.0 inches. Pipe I.D. = 4.0 in.

12-3   The flow of kerosene is being measured with an orifice meter similar to that shown in Fig. 12-5. The pipe is a 2-inch Schedule 40 pipe and the orifice diameter is 1.00 inch. The kerosene is at 77°F. For a pressure difference of 0.53 psi across the orifice, calculate the volume flow rate of kerosene.

12-4   A sharp-edged orifice meter is placed in a 10-inch diameter pipe carrying ammonia. If the volume flow rate is 25 gal/min calculate the deflection of a water manometer (a) if the orifice diameter is 1.0 inch, and (b) if the orifice diameter is 7.0 inches. The ammonia has a specific gravity of 0.83 and a dynamic viscosity of 2.5×10^{-6} lb-sec/ft^2.

12-5 A pitot-static tube is inserted into a pipe carrying methyl alcohol at 77°F. A differential manometer using mercury as the gage fluid is connected to the tube and shows a deflection of 8.8 inches. Calculate the velocity of flow of the alcohol.

12-6 A pitot-static tube is connected to a differential manometer using water at 100°F as the gage fluid. The velocity of air at 100°F and atmospheric pressure is to be measured and it is expected that the maximum velocity will be 5000 ft/min. Calculate the expected manometer deflection.

12-7 A pitot-static tube is inserted in a pipe carrying water at 50°F. A differential manometer using mercury as the gage fluid shows a deflection of 4.2 inches. Calculate the velocity of flow.

12-8 Determine the maximum possible flow rate over a 60° V-notch weir if the width of the notch at the top is 12 inches.

12-9 Determine the required length of a contracted weir similar to that shown in Fig. 12-17(b) to pass 15 ft^3/sec of water. The height of the crest is to be 3 feet from the channel bottom and the maximum head above the crest is to be 18 inches.

12-10 Plot a graph of Q versus H for a full width rectangular weir with a crest length of 6 feet and whose crest is 2 feet from the channel bottom. Consider values of the head H from zero to 12 inches in 2-inch steps.

12-11 On the same graph as that used for problem 12-10 plot the curve of Q versus H for a weir having the same dimensions except that it is placed in a channel wider than 6 feet. It thus becomes a contracted weir.

12-12 Compare the discharges over the following weirs when the head H is 18 inches.
(a) Full width rectangular: $L = 3$ ft, $H_c = 4$ ft
(b) Contracted rectangular: $L = 3$ ft, $H_c = 4$ ft
(c) 90° V-notch (top width also 3 feet)

12-13 Plot a graph of Q versus H for a 90° V-notch weir for values of the head from zero to 12 inches in 2-inch steps.

12–14 For a Parshall flume having a throat width of 9 inches, calculate the head H corresponding to the minimum and maximum flows.

12–15 For a Parshall flume having a throat width of 8 feet, calculate the head H corresponding to the minimum and maximum flows. Plot a graph of Q versus H using five values of H spaced approximately equally between the minimum and maximum.

12–16 A flow rate of 50 ft³/sec falls within the range of both the 4 feet and the 10 feet wide Parshall flume. Compare the head H for this flow rate in each size.

13

Forces Due to
Fluids in Motion

Force Equation

Whenever the magnitude or direction of the velocity of a body is changed, a force is required to accomplish the change. Newton's second law of motion is often used to express this concept in mathematical form, the most common form being

$$F = ma \qquad (13\text{-}1)$$

Force equals mass times acceleration. Acceleration is the time rate of change of velocity. However, since velocity is a vector quantity having both magnitude and direction, changing either the magnitude or the direction would result in an acceleration. According to Eq. (13-1), an external force would be required to cause the change.

Examples in fluid mechanics in which an external force is required to cause a change in the velocity of the flow of fluids are numerous. The nozzle of a fire hose shooting a high velocity stream of water must be firmly held or it will thrash wildly due to the force of the water on the nozzle. Wind striking a flat sign exerts a force on the sign because the direction of the air flow is being changed. High velocity water or steam striking the buckets of a turbine exert a force causing the turbine to rotate and generate power. In piping installations, forces are exerted on elbows where the direction of fluid flow is changed, and on expansions and contractions where the velocity is changed. These elements must be firmly anchored to resist these forces, especially if high velocity flows or rapid surges of flow are experienced.

Equation (13-1) is convenient for use with solid bodies since the mass remains constant and the acceleration of the entire body can be determined. In fluid flow problems, a continuous flow is caused to undergo the acceleration and a different form of Newton's equation is desirable. Because acceleration is the time rate of change of velocity, Eq. (13–1) can be written as

$$F = ma = m\frac{\Delta v}{\Delta t} \tag{13-2}$$

The term $m/\Delta t$ can be interpreted as the mass flow rate, that is, the amount of mass flowing in a given amount of time. In the discussion of fluid properties earlier in this book, mass flow rate was indicated by the symbol M. Also, M is related to the volume flow rate Q by the relationship

$$M = \rho Q \tag{13-3}$$

where ρ is the density of the fluid. Then Eq. (13–2) becomes

$$F = (m/\Delta t)\, \Delta v = M\Delta v = \rho Q\Delta v \tag{13-4}$$

This is the general form of the force equation for use in fluid flow problems since it involves the velocity and volume flow rate, items generally known in a fluid flow system.

It must be emphasized that problems involving forces must account for the directions in which the forces act. In Eq. (13–4), force and velocity are both vector quantities. The equation is valid only when all terms have the same direction. For this reason, different equations are written in each direction of concern in a particular case. In general, if three perpendicular directions are called x, y, and z, a separate equation can be written for each direction.

$$F_x = \rho Q\Delta v_x = \rho Q(v_{2x} - v_{1x}) \tag{13-5}$$
$$F_y = \rho Q\Delta v_y = \rho Q(v_{2y} - v_{1y}) \tag{13-6}$$
$$F_z = \rho Q\Delta v_z = \rho Q(v_{2z} - v_{1z}) \tag{13-7}$$

This is the form of the force equation which will be used in this book, with the directions chosen according to the physical situation. In a particular direction, say x, the term F_x refers to the net external force which acts on the fluid in that direction. Therefore, it is the algebraic sum of *all* external forces including that exerted by a solid surface and forces due to fluid pressure. The term Δv_x refers to the change in velocity in the x direction. Also, v_1 is the velocity as the fluid enters the device and v_2 is the velocity as it leaves. Then v_{1x} is the component of v_1 in the x direction and v_{2x} is the component of v_2 in the x direction. The example problems presented in the following sections will illustrate the approach to problems of this type.

Equation (13–4) is often called the impulse-momentum equation because the product of mass times a change of velocity, $m\Delta v$, is the change in momentum. Since we will be dealing primarily with forces produced by fluids, the name "force equation" is used.

13–2 Forces on Bends in Pipe Lines

Figure 13–1 shows a typical $90°$ elbow in a pipe carrying a steady volume flow rate Q. In order to ensure proper installation, it is important to know how much force is required to hold it in equilibrium. The following problem demonstrates an approach to this type of situation.

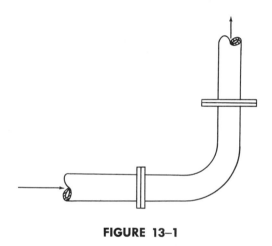

FIGURE 13–1

Example Problem 13–1. Calculate the force which must be exerted on the pipe elbow shown in Fig. 13–1 to hold it in equilibrium. The elbow is in a horizontal plane and is connected to two 4-inch Schedule 40 pipes carrying 800 gal/min of water at $60°F$. The inlet pressure is 80 psig.

Solution. The problem may be visualized by considering the fluid within the elbow to be a free body as shown in Fig. 13–2. Forces are shown as solid vectors while the direction of the velocity of flow is shown by dashed vectors. A convention must be set for the directions of all vectors. Here it is assumed that the positive x direction is to the left and the positive y direction is up. The forces R_x and R_y are the external reactions required to maintain equilibrium. The forces $p_1 A_1$ and $p_2 A_2$ are the forces due to the fluid pressure. The two directions will be analyzed separately.

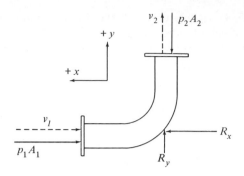

FIGURE 13–2

x direction:

$$F_x = \rho Q(v_{2x} - v_{1x})$$

But,

$$F_x = R_x - p_1 A_1$$
$$v_{2x} = 0$$
$$v_{1x} = -v_1$$

Then,

$$R_x - p_1 A_1 = \rho Q[0 - (-v_1)]$$
$$R_x = \rho Q v_1 + p_1 A_1 \tag{13–8}$$

From the given data $p_1 = 80$ psig, $\rho = 1.94$ slugs/ft³ $= 1.94$ lb-sec²/ ft⁴, and $A_1 = 12.73$ in.² $= 0.0884$ ft². Then,

$$Q = 800 \text{ gal/min} \times \frac{1 \text{ ft}^3/\text{sec}}{449 \text{ gal/min}} = 1.78 \text{ ft}^3/\text{sec}$$

$$v_1 = \frac{Q}{A_1} = \frac{1.78 \text{ ft}^3/\text{sec}}{0.0884 \text{ ft}^2} = 20.2 \text{ ft/sec}$$

$$\rho Q v_1 = 1.94 \frac{\text{lb-sec}^2}{\text{ft}^4} \cdot \frac{1.78 \text{ ft}^3}{\text{sec}} \cdot \frac{20.2 \text{ ft}}{\text{sec}} = 69.8 \text{ lb}$$

$$p_1 A_1 = \frac{80 \text{ lb}}{\text{in.}^2} \cdot 12.73 \text{ in.}^2 = 1020 \text{ lb}$$

Substituting these values into Eq. (13–8) gives

$$R_x = (69.8 + 1020) \text{ lb} = 1090 \text{ lb}$$

y direction:

$$F_y = \rho Q(v_{2y} - v_{1y})$$

But,

$$F_y = R_y - p_2A_2$$
$$v_{2y} = +v_2$$
$$v_{1y} = 0$$

Then,

$$R_x - p_2A_2 = \rho Q v_2$$
$$R_x = \rho Q v_2 + p_2A_2$$

If energy losses in the elbow are neglected, $v_2 = v_1$ and $p_2 = p_1$ since the sizes of the inlet and outlet are equal. Then,

$$\rho Q v_2 = 69.8 \text{ lb}$$
$$p_2A_2 = 1020 \text{ lb}$$
$$R_y = (69.8 + 1020) \text{ lb} = 1090 \text{ lb}$$

The forces R_x and R_y are the reactions caused at the elbow as the fluid turns 90°. They may be supplied by anchors on the elbow or taken up through the flanges into the main pipes.

Example Problem 13–2. Linseed oil, having a specific gravity of 0.93, enters the reducing bend shown in Fig. 13–3 with a velocity of 10 ft/sec and a pressure of 40 psig. The bend is in a horizontal plane. Calculate the x and y forces required to hold the bend in place. Neglect energy losses in the bend.

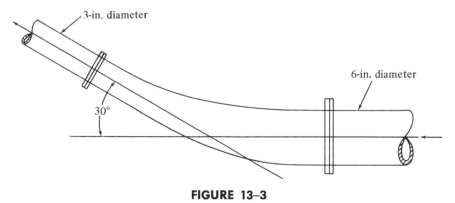

3-in. diameter

6-in. diameter

30°

FIGURE 13–3

Solution. The fluid in the bend is shown as a free body in Fig. 13–4. The force equations for the x and y directions shown are developed below.

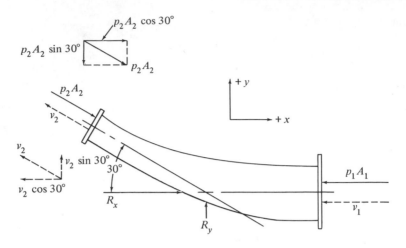

FIGURE 13–4

x direction:

$$F_x = \rho Q(v_{2x} - v_{1x})$$
$$R_x - p_1 A_1 + p_2 A_2 \cos 30° = \rho Q[-v_2 \cos 30° - (-v_1)]$$
$$R_x = p_1 A_1 - p_2 A_2 \cos 30° - \rho Q v_2 \cos 30° + \rho Q v_1 \qquad \textbf{(13–9)}$$

Algebraic signs must be carefully included according to the sign convention shown in Fig. 13–4. Notice that all forces and velocity terms are the components *in the* x *direction*.

y direction:

$$F_y = \rho Q(v_{2y} - v_{1y})$$
$$R_y - p_2 A_2 \sin 30° = \rho Q(v_2 \sin 30° - 0)$$
$$R_y = p_2 A_2 \sin 30° + \rho Q v_2 \sin 30° \qquad \textbf{(13–10)}$$

The numerical values of several items must now be calculated. From the appendix, $A_1 = 28.27$ in.$^2 = 0.1964$ ft^2 and $A_2 = 7.069$ in.$^2 = 0.0491$ ft^2.

$$\rho = (sg)\,(\rho_w) = (0.93)(1.94 \text{ slugs/ft}^3)$$
$$\rho = 1.80 \text{ slugs/ft}^3 = 1.80 \text{ lb-sec}^2/\text{ft}^4$$
$$Q = A_1 v_1 = (0.1964 \text{ ft}^2)(10 \text{ ft/sec}) = 1.964 \text{ ft}^3/\text{sec}$$

Since $A_1 v_1 = A_2 v_2$ because of continuity,

$$v_2 = v_1(A_1/A_2) = (10 \text{ ft/sec})(0.1964/0.0491) = 40 \text{ ft/sec}$$

Bernoulli's equation can be used to find p_2.

$$\frac{p_1}{\gamma} + z_1 + \frac{v_1^2}{2g} = \frac{p_2}{\gamma} + z_2 + \frac{v_2^2}{2g}$$

But $z_1 = z_2$. Then,

$$p_2 = p_1 + \gamma(v_1{}^2 - v_2{}^2)/2g$$
$$= 40 \text{ psig} + \frac{(0.93)(62.4)(100 - 1600)}{(2)(32.2)}\frac{\text{lb}}{\text{ft}^2} \cdot \frac{1 \text{ ft}^2}{144 \text{ in.}^2}$$
$$p_2 = 40 \text{ psig} - 9.4 \text{ psi} = 30.6 \text{ psig}$$

The quantities needed for Eqs. (13–9) and (13–10) are

$$p_1 A_1 = (40 \text{ lb/in.}^2)(28.27 \text{ in.}^2) = 1130 \text{ lb}$$
$$p_2 A_2 = (30.6 \text{ lb/in.}^2)(7.069 \text{ in.}^2) = 216 \text{ lb}$$
$$\rho Q v_1 = (1.80)(1.964)(10) \text{ lb} = 35.3 \text{ lb}$$
$$\rho Q v_2 = (1.80)(1.964)(40) \text{ lb} = 141.5 \text{ lb}$$

From Eq. (13–9),

$$R_x = (1130 - 216 \cos 30° - 141.5 \cos 30° + 35.3) \text{ lb}$$
$$R_x = 855 \text{ lb}$$

From Eq. (13–10),

$$R_y = 216 \sin 30° + 141.5 \sin 30°$$
$$R_y = 179 \text{ lb}$$

13–3   Forces on Stationary Objects

When free streams of fluid are deflected by stationary objects, there must be external forces exerted to maintain the object in equilibrium. Some examples are described below.

Example Problem 13–3. A jet of water having a velocity of 20 ft/sec is deflected by a curved vane 90° as shown in Fig. 13–5. If the jet is one

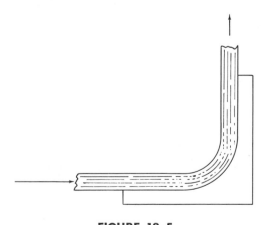

FIGURE 13–5

inch in diameter and flowing freely in the atmosphere, calculate the x and y forces exerted on the water by the vane.

Solution. For the x direction, using the force diagram of Fig. 13–6,

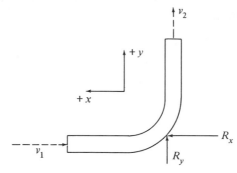

FIGURE 13–6

$$F_x = \rho Q(v_{2x} - v_{1x})$$
$$R_x = \rho Q[0 - (-v_1)] = \rho Q v_1$$

But,

$$Q = Av = (0.00545 \text{ ft}^2)(20 \text{ ft/sec}) = 0.109 \text{ ft}^3/\text{sec}$$

Then,

$$R_x = (1.94)(0.109)(20) \text{ lb} = 4.23 \text{ lb}$$

For the y direction, assuming $v_2 = v_1$,

$$F_y = \rho Q(v_{2y} - v_{1y})$$
$$R_y = \rho Q(v_2 - 0)$$
$$R_y = (1.94)(0.109)(20) \text{ lb} = 4.23 \text{ lb}$$

Example Problem 13–4. In a decorative fountain, 1.5 ft³/sec of water having a velocity of 25 ft/sec is being deflected by the angled chute shown in Fig. 13–7. Determine the reactions on the chute in the x and y directions shown. Also calculate the total resultant force and the direction in which it acts.

Solution. The reaction in the x direction is assumed to act toward the left while the reaction in the y direction is assumed to act upward. These directions are taken as positive. Figure 13–8 shows the velocity vectors broken into components in the x and y directions. The force equation in the x direction is developed below.

$$F_x = \rho Q(v_{2x} - v_{1x})$$

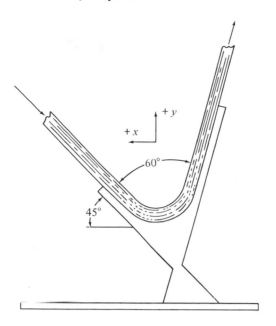

FIGURE 13-7

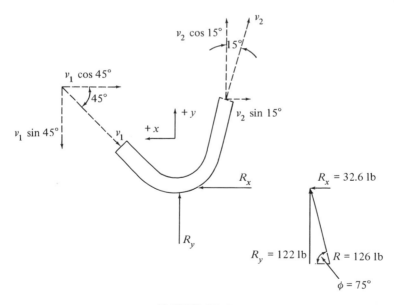

FIGURE 13-8

But,

$$v_{2x} = -v_2 \sin 15° \qquad \text{(toward the right)}$$
$$v_{1x} = -v_1 \cos 45° \qquad \text{(toward the right)}$$

Neglecting friction in the chute, it can be assumed that $v_2 = v_1$. Also, the only external force is R_x. Then,

$$R_x = \rho Q[-v_2 \sin 15° - (-v_1 \cos 45°)]$$
$$R_x = \rho Q v(-\sin 15° + \cos 45°)$$

Using $\rho = 1.94$ slugs/ft^3 for water,

$$R_x = (1.94)(1.50)(25)(-0.259 + 0.707) \text{ lb}$$
$$R_x = 32.6 \text{ lb}$$

In the y direction,

$$F_y = \rho Q(v_{2y} - v_{1y})$$

But,

$$v_{2y} = v_2 \cos 15° \qquad \text{(upward)}$$
$$v_{1y} = -v_1 \sin 45° \text{ (downward)}$$

Then,

$$R_y = \rho Q[v_2 \cos 15° - (-v_1 \sin 45°)]$$
$$R_y = \rho Q v(\cos 15° + \sin 45°)$$
$$R_y = (1.94)(1.5)(25)(0.966 + 0.707) \text{ lb}$$
$$R_y = 122 \text{ lb}$$

For the resultant force R,

$$R = \sqrt{R_x^2 + R_y^2} = \sqrt{(32.6)^2 + (122)^2}$$
$$R = 126 \text{ lb}$$

For the direction of R,

$$\tan \phi = R_y/R_x = 122/32.6 = 3.74$$
$$\phi = 75.0°$$

Therefore, the resultant force which the chute must exert on the water is 126 lb acting 75° from the horizontal as shown in Fig. 13–8.

Example Problem 13–5. When a stream of a fluid strikes a flat surface, the stream splits into two or more parts depending on the geometry of the jet and the surface. An example is shown in Fig. 13–9 in which the stream is a long sheet of cooling oil flowing from a slotted nozzle. The side view shows the horizontal stream striking the flat plate at an angle. The nozzle opening is 1 inch wide and 3.5 feet long and the velocity of the stream is 20 ft/sec. The cooling oil has a specific gravity of 0.90. If the plate is inclined at an angle of 50° with the horizontal, determine how the stream splits and the total force on the plate.

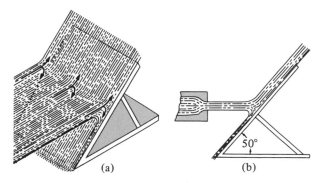

(a)   (b)

FIGURE 13–9

Solution. Since the fluid stream splits, there are two paths which the fluid leaving the plate may take. The force equations, as used in preceding examples and stated in Eqs. (13–5), (13–6), and (13–7), are not convenient to use here because they were written assuming the flow to be a single continuous stream. However, slightly different equations can be developed. Consider a particular direction, say x. We can write

$$F_x = (\rho Q v_x)_{\text{leaving}} - (\rho Q v_x)_{\text{entering}} \qquad (13\text{–}11)$$

The product $\rho Q v$ is referred to as momentum flux and can be interpreted as the rate of transfer of momentum. Then Eq. (13–11) expresses the concept that the net external force on a body of fluid in a particular direction is equal to the net change in momentum flux in that direction. Either the leaving or entering momentum flux could be composed of the sum of the momentum flux of several individual fluid streams. This concept will be applied to the solution of the problem at hand.

Figure 13–10 represents a side view of the plate. The convenient directions for use in the analysis are perpendicular and parallel to the plate, called x and y respectively in the figure. The fluid inside the box made with dashed lines is considered the free body.

Then, in the x direction, Eq. (13–11) applies where,

$$F_x = R_x$$
$$(\rho Q v_x)_{\text{leaving}} = 0$$
$$(\rho Q v_x)_{\text{entering}} = -\rho Q_1 v_1 \sin \theta$$

Notice that none of the fluid leaving the plate flows in the x direction, and only a component of the entering fluid does. Then,

$$R_x = 0 - (-\rho Q_1 v_1 \sin \theta) = \rho Q_1 v_1 \sin \theta$$

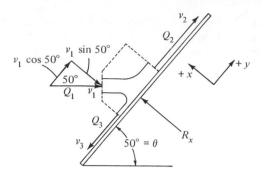

FIGURE 13-10

But,

$$\rho = (0.90)(1.94 \text{ slugs/ft}^3) = 1.746 \text{ slugs/ft}^3$$

$$Q_1 = A_1 v_1 = (1 \text{ in.})(3.5 \text{ ft}) \cdot \frac{20 \text{ ft}}{\text{sec}} \cdot \frac{1 \text{ ft}}{12 \text{ in.}}$$

$$Q_1 = 5.84 \text{ ft}^3/\text{sec}$$

Then,

$$R_x = (1.746)(5.84)(20)(\sin 50°) = 156 \text{ lb}$$

The manner in which the flow splits can be calculated by considering the momentum flux change in the y direction.

$$F_y = (\rho Q v_y)_{\text{leaving}} - (\rho Q v_y)_{\text{entering}} \qquad (13\text{–}12)$$

The only possible force along the surface of the plate is that due to friction which is quite small. Therefore it is assumed that $F_y = 0$. The entering flow is the single stream Q_1. However, the leaving flow is composed of Q_2 flowing up the plate and Q_3 flowing down the plate. Then,

$$(\rho Q v_y)_{\text{leaving}} = \rho Q_2 v_2 - \rho Q_3 v_3$$

$$(\rho Q v_y)_{\text{entering}} = \rho Q_1 v_1 \cos \theta$$

Equation (13–12) becomes

$$0 = \rho Q_2 v_2 - \rho Q_3 v_3 - \rho Q_1 v_1 \cos \theta \qquad (13\text{–}13)$$

If friction is neglected, $v_1 = v_2 = v_3$. These and the density ρ can be cancelled from the equation, which then becomes

$$0 = Q_2 - Q_3 - Q_1 \cos \theta \qquad (13\text{–}14)$$

Because of continuity,

$$Q_1 = Q_2 + Q_3$$

$$Q_3 = Q_1 - Q_2 \qquad (13\text{–}15)$$

Substituting this into Eq. (13–14),

$$0 = Q_2 - (Q_1 - Q_2) - Q_1 \cos \theta$$
$$0 = Q_2 - Q_1 + Q_2 - Q_1 \cos \theta$$
$$0 = 2Q_2 - Q_1(1 + \cos \theta)$$

Solving for Q_2,

$$Q_2 = \tfrac{1}{2} Q_1(1 + \cos \theta) \qquad\qquad (13\text{–}16)$$

Equation (13–15) can be used to calculate Q_3.

$$Q_3 = Q_1 - Q_2 = Q_1 - \tfrac{1}{2} Q_1(1 + \cos \theta)$$
$$Q_3 = \tfrac{1}{2} Q_1(1 - \cos \theta) \qquad\qquad (13\text{–}17)$$

Evaluating Eqs. (13–16) and (13–17) for $\theta = 50°$,

$$Q_2 = (0.50)(5.84 \text{ ft}^3/\text{sec})(1 + \cos 50°) = 4.80 \text{ ft}^3/\text{sec}$$
$$Q_3 = (0.50)(5.84 \text{ ft}^3/\text{sec})(1 - \cos 50°) = 1.04 \text{ ft}^3/\text{sec}$$

13–4   **Forces on Moving Objects**

The vanes of turbines and other rotating machinery are familiar examples of moving objects which are acted upon by high velocity fluids. A jet of fluid, having a velocity greater than that of the blades of the turbine, exerts a force on the blades causing them to accelerate or to generate useful mechanical energy. When dealing with forces on moving bodies, the *relative motion* of the fluid with respect to the body must be considered.

Example Problem 13–6.   Figure 13–11(a) shows a jet of water having a velocity v_1 striking a vane which is moving with a velocity v_0. Determine the forces exerted by the vane on the water if $v_1 = 60$ ft/sec and $v_0 = 25$ ft/sec. The jet is 1.50 inches in diameter.

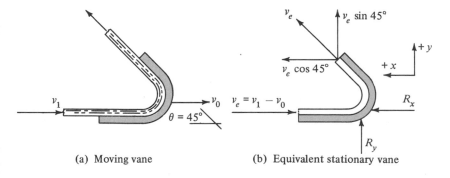

(a) Moving vane   (b) Equivalent stationary vane

FIGURE 13–11

Solution. The system with a moving vane can be converted into an equivalent stationary system as shown in Fig. 13–11(b) by defining an effective velocity v_e and an effective volume flow rate Q_e.

$$v_e = v_1 - v_0 \tag{13–18}$$
$$Q_e = A_1 v_e \tag{13–19}$$

where A_1 is the area of the jet as it enters the vane. It is only the difference between the jet velocity and the vane velocity which is effective in creating a force on the vane. The force equations can be written in terms of v_e and Q_e. In the x direction,

$$R_x = \rho Q_e v_e \cos \theta - (-\rho Q_e v_e)$$
$$R_x = \rho Q_e v_e (1 + \cos \theta) \tag{13–20}$$

In the y direction,

$$R_y = \rho Q_e v_e \sin \theta - 0 \tag{13–21}$$

But,

$$v_e = v_1 - v_0 = (60 - 25) \text{ ft/sec} = 35 \text{ ft/sec}$$
$$Q_e = A_1 v_e = (0.01227 \text{ ft}^2)(35 \text{ ft/sec}) = 0.429 \text{ ft}^3/\text{sec}$$

Then the reactions are calculated from Eqs. (13–20) and (13–21).

$$R_x = (1.94)(0.429)(35)(1 + \cos 45°) = 49.7 \text{ lb}$$
$$R_y = (1.94)(0.429)(35)(\sin 45°) = 20.6 \text{ lb}$$

13–5   **Jet Propulsion**

An aircraft or boat is propelled by jet propulsion when a stream of fluid (a jet) is discharged from the craft at a higher velocity than the velocity of the craft itself. In an aircraft jet engine, sketched in Fig. 13–12, air is taken into a compressor at a velocity essentially equal to that of the aircraft, v_1. After being compressed it is mixed with fuel and burned in a combustion chamber. The hot gas products of com-

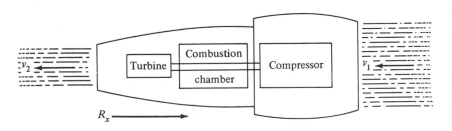

FIGURE 13–12

bustion pass through a turbine which takes some energy to drive the compressor. The gases then exit from the rear of the engine at a high velocity v_2. The action of the hot gases passing from the engine causes a reaction on the aircraft propelling it forward. The total momentum flux change is the sum of the change for the air flowing through the engine and for the fuel burned in the combustion chamber and then exhausted. The net reaction on the aircraft is then

$$R_x = \rho_a Q_a(v_2 - v_1) + \rho_f Q_f v_2 \qquad \qquad \textbf{(13–22)}$$

The subscript a refers to the air and f refers to the fuel. Note that the velocities v_2 and v_1 are measured relative to the aircraft and that the initial velocity of the fuel relative to the aircraft is zero. The reaction R_x is usually referred to as the thrust of the engine.

13–6   Rocket Propulsion

The thrust produced by a rocket engine is also due to the discharge of a high velocity stream of fluid from the rear of the engine. One important difference between rocket engines and jet engines is that a rocket propelled craft carries both the fuel and the oxidizer to cause combustion while a jet engine must take in air from outside. Figure 13–13 shows a sketch of a rocket engine. The fuel and oxidizer are brought together in

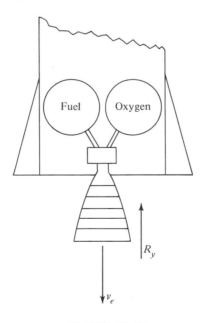

FIGURE 13–13

the combustion chamber and ignited. The hot gases pass through a nozzle and are exhausted at a high velocity v_e relative to the engine. Since both the fuel and oxidizer are initially at rest relative to the engine, the reaction is equal to the leaving momentum flux of the gases.

$$R_y = \rho Q v_e \qquad (13\text{–}23)$$

The product ρQ is usually called the mass flow rate M in slugs/sec since this indicates the rate at which fuel is used by the engine.

PRACTICE PROBLEMS

13–1   Calculate the force required to hold a 90° elbow in place when attached to 6-inch Schedule 40 pipes carrying water at 4.45 ft³/sec and 150 psig.

13–2   Calculate the force required to hold a flat plate in equilibrium perpendicular to the flow of water at 75 ft/sec issuing from a 3-inch diameter nozzle.

13–3   What must be the velocity of flow of water from a 2-inch diameter nozzle to exert a force of 300 pounds on a flat wall?

13–4   Calculate the force exerted on a stationary curved vane which deflects a 1-inch diameter stream of water through a 90° angle. The volume flow rate is 150 gal/min.

13–5   Calculate the force on a 45° elbow attached to an 8-inch Schedule 80 steel pipe carrying water at 80°F at 6.5 ft³/sec. The outlet of the elbow discharges into the atmosphere. Consider the energy loss in the elbow.

13–6   A highway sign is being designed to withstand winds of 80 miles per hour. Calculate the total force on a sign 12 ft × 10 ft if the wind is flowing perpendicular to the face of the sign. Calculate the equivalent pressure on the sign in lb/ft². The air is at 40°F. (See Chapter 14 and problem 14–9 for a more complete analysis of this problem.)

13–7   Calculate the force required to hold a 180° close return bend in equilibrium. The bend is in a horizontal plane and is attached to a 4-inch Schedule 160 steel pipe carrying 500 gal/min of a hydraulic fluid at 3000 psi. The fluid has a specific gravity of 0.89.

13–8   A bend in a tube causes the flow to turn through an angle of 135°. The pressure ahead of the bend is 40 psig. If the 6-inch Type L copper tube carries 4 ft³/sec of carbon tetrachloride at 77°F, determine the force on the bend.

13–9   A 2-inch diameter stream of water having a velocity of 40 ft/sec strikes the edge of a flat plate such that half the stream is deflected downward as shown in Fig. 13–14. Calculate the force on the plate and the moment due to the force at point A.

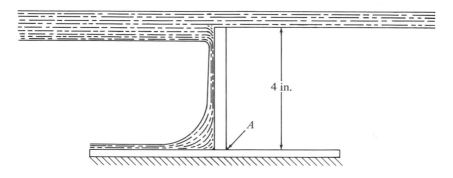

4 in.

A

FIGURE 13–14

13–10 A stream of water flows onto a flat plate as shown in Fig. 13–15. Plot a graph of the flow leaving the plate to the right, Q_2, versus the angle θ if $Q_1 = 1.0$ ft³/sec. Also plot the magnitude of the force on the plate in the vertical direction versus θ. Use 15° increments on the angle from 0° to 180°.

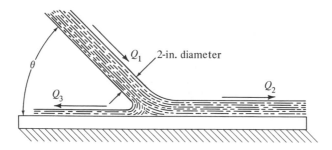

θ

Q_1 2-in. diameter

Q_2

Q_3

FIGURE 13–15

13–11 Sea water (sg = 1.03) enters a heat exchanger through a reducing bend connecting a 4-inch Type K copper tube with a 2-inch Type K tube. The pressure upstream from the bend is 120 psig.

Calculate the force required to hold the bend in equilibrium. Consider the energy loss in the bend assuming it has a loss coefficient C_L of 3.5. The flow rate is 0.8 ft^3/sec.

13–12   A reducer connects a standard 6-inch Schedule 40 pipe to a 3-inch Schedule 40 pipe. The walls of the conical reducer are tapered at an included angle of 40°. The flow rate of water is 500 gal/min and the pressure ahead of the reducer is 125 psig. Considering the energy loss in the reducer, calculate the force exerted on the reducer by the water.

13–13   Water is piped vertically from below a boat and discharged horizontally in a jet 4 inches in diameter with a velocity of 60 ft/sec. Calculate the force on the boat.

13–14   A 2-inch nozzle is attached to a hose having an inside diameter of 4 inches. The loss coefficient C_L of the nozzle is 0.12 based on the outlet velocity head. If the jet issuing from the nozzle has a velocity of 80 ft/sec calculate the force exerted by the water on the nozzle.

13–15   A flat plate weighing 10 pounds rests on a surface as shown in Fig. 13–16. A splitter causes $\frac{3}{4}$ of the stream of water to flow to the left while $\frac{1}{4}$ flows to the right. The stream is 3 inches in diameter and has a velocity of 40 ft/sec. If the coefficient of friction between the plate and the fixed surface is 0.30, determine if the plate will slide.

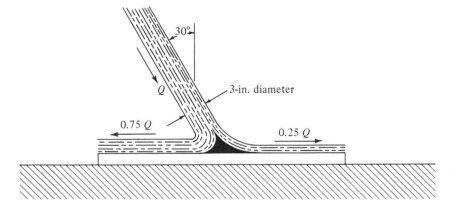

FIGURE 13–16

13–16   For the plate in problem 13–15, what must be the percentage of flow to the left for the motion of the plate to be impending?

13–17   Figure 13–17 represents a type of flow meter in which the flat vane is rotated on a pivot as it deflects the fluid stream. The fluid force is counterbalanced by a spring. Calculate the spring force required to hold the vane in a vertical position when water at 100 gal/min flows from the 1-inch Schedule 40 pipe to which the meter is attached.

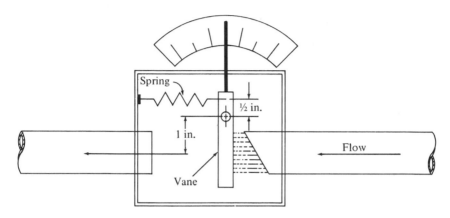

FIGURE 13–17

13–18   A vehicle is to be propelled by a jet of water impinging on a vane as shown in Fig. 13–18. The jet has a velocity of 100 ft/sec and an area of 36 in.² Calculate the force on the vehicle (a) if it is stationary, (b) if it is moving at 40 ft/sec.

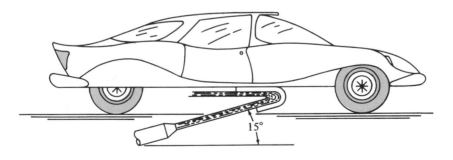

FIGURE 13–18

13–19   A jet engine is tested at standard atmospheric conditions (see Table 14–2) on a stationary test stand. Air enters the engine at the rate of 5000 ft³/sec. Fuel is added at the rate of 1 pound of fuel for each 100 pounds of air. The products of combustion leave the engine at 750 ft/sec. Calculate the thrust produced by the engine.

13–20 If the engine of Problem 13–19 is operating at 450 miles per hour at an elevation of 20,000 ft in standard atmosphere, calculate the thrust. See Table 14–3 for the properties of air.

13–21 A rocket engine discharges 40 lb/sec of combustion gases at 4000 ft/sec. Calculate the thrust produced by the engine.

13–22 The first stage of a space launch vehicle has five rocket engines which burn 15 tons of fuel and oxidizer per second. If a total thrust of 7.5 million pounds is produced, calculate the exit velocity of the burned gases.

14

Forces on Immersed Bodies

14–1   Drag and Lift

A moving body immersed in a fluid experiences forces caused by the action of the fluid. The total effect of these forces is quite complex. However, for the purposes of design or the analysis of the behavior of a body in a fluid, two resultant forces, drag and lift, are the most important. Lift and drag forces are the same regardless of whether the body is moving in the fluid or the fluid is moving over the body.

Drag is the force on a body caused by the fluid which resists motion in the direction of travel of the body. The most familiar applications requiring the study of drag are in the transportation fields. "Wind resistance" is the term often used to describe the effects of drag on aircraft, automobiles, trucks, and trains. The drag force must be opposed by a propulsive force in the opposite direction in order to maintain or increase the velocity of the vehicle. Since the production of the propulsive force requires added power, it is desirable to minimize drag.

Lift is a force caused by the fluid in a direction perpendicular to the direction of travel of the body. Its most important application is in the design and analysis of aircraft wings called airfoils. The geometry of an airfoil is such that a lift force is produced as air passes over and under it. Of course, the magnitude of the lift must at least equal the weight of the aircraft in order for it to fly.

The study of the performance of bodies in moving air streams is called aerodynamics. Gases other than air could be considered in this field but due to the obvious importance of the applications in aircraft design, the great majority of work has been done with air as the fluid.

Hydrodynamics is the name given to the study of moving bodies immersed in liquids, particularly water. Many concepts concerning lift and drag are similar regardless of whether the fluid is a liquid or a gas. This is not true, however, at high velocities where the effects of the compressibility of the fluid must be taken into account. Liquids can be considered incompressible in the study of lift and drag. Conversely, a gas such as air is readily compressible.

Much of the practical data concerning lift and drag has been generated experimentally. Some of these data will be reported here to illustrate the concepts. The references listed at the end of this chapter include more comprehensive treatments of the subject.

14–2   Coefficient of Drag

Drag forces are usually expressed in the form

$$\text{Drag} = C_D(\tfrac{1}{2}\rho v^2)A \qquad\qquad (14–1)$$

The term C_D is the drag coefficient, a dimensionless factor whose magnitude depends primarily on the physical shape of the object and its orientation relative to the fluid stream. The quantity $\tfrac{1}{2}\rho v^2$ is the dynamic pressure as defined below. The term A refers to a characteristic area of the body, either the surface area or the maximum cross sectional area perpendicular to the direction of flow as discussed in Sections 14–3, 14–4, and 14–6.

The influence of the dynamic pressure on drag can be visualized by referring to Fig. 14–1 which shows a cylinder in a fluid stream. The streamlines depict the path of the fluid as it approaches and flows around the cylinder. At point s on the surface of the cylinder, the fluid stream is at rest or "stagnant." The term *stagnation point* is used to describe this point. The relationship between the pressure p_s and that in the undisturbed stream at point 1 can be found using Bernoulli's equation.

$$\frac{p_1}{\gamma} + \frac{v_1^2}{2g} = \frac{p_s}{\gamma} \qquad\qquad (14–2)$$

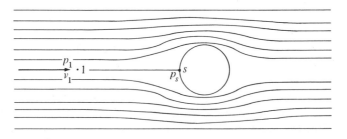

FIGURE 14–1

Solving for p_s,

$$p_s = p_1 + \frac{\gamma v_1^2}{2g}$$

But since $\rho = \gamma/g$,

$$p_s = p_1 + \tfrac{1}{2}\rho v_1^2 \qquad\qquad (14\text{–}3)$$

Then the stagnation pressure is greater than the static pressure in the free stream by the magnitude of the dynamic pressure $\tfrac{1}{2}\rho v_1^2$. The kinetic energy of the moving stream is transformed into a kind of potential energy in the form of pressure.

The increase in pressure at the stagnation point can be expected to produce a force on the body opposing its motion, that is, a drag force. However, the magnitude of the force is dependent not only on the stagnation pressure but also on the pressure at the back side of the body. Since it is impossible to predict the actual variation in pressure on the back side, the drag coefficient is used.

The total drag on a body is due to two components.* *Pressure drag* (also called form drag) is due to the disturbance of the flow stream as it passes the body creating a turbulent wake. The characteristics of the disturbance are dependent on the form of the body and sometimes on the Reynolds number of flow and the roughness of the surface. *Friction drag* is due to shearing stresses in the thin layer of fluid near the surface of the body called the boundary layer. These two types of drag are described in the sections that follow.

14–3   **Friction Drag**

The methods used to determine the friction drag on the surface of a body are conventionally based on equations developed for flat plates moving parallel to the fluid stream. Figure 14–2 shows such a plate. Beginning at the leading edge of the plate, a layer of fluid near the surface of the plate has its velocity changed to a value below that of the approaching free stream. The thickness of the affected layer, called the boundary layer, increases with increasing distance from the edge of the plate. For a relatively long plate the layer is first laminar, then becomes transitional, and finally becomes turbulent. The point at which the transition starts is dependent on the Reynolds number and also the roughness of the surface. A body with a very rough surface near the front will set

---

*For a lifting body such as an airfoil a third component exists as described in Section 14–6.

up disturbances in the boundary layer which cause the laminar flow to be unstable and change to turbulent flow at low Reynolds numbers. The value of the coefficient of friction drag is dependent on whether the flow is laminar, transitional, or turbulent.

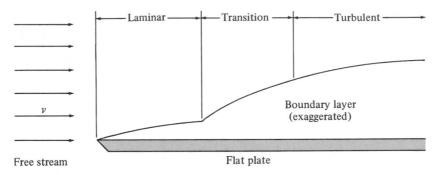

FIGURE 14–2

The method of computing the Reynolds number for an immersed body is different from that used for closed pipes and open channels. A general form for the Reynolds number is

$$N_R = \frac{\rho v L}{\mu} = \frac{v L}{\nu} \tag{14-4}$$

As before, ρ is the fluid density, μ is the dynamic viscosity, and ν is the kinematic viscosity of the fluid. However, the velocity v is now taken as the velocity of the undisturbed fluid upstream from the body instead of the average velocity. The term L is some characteristic length of the body in question. It is very important to understand which dimension is taken for L in the calculation of the Reynolds number for a particular application. In the case of flow over a flat surface, L is the length of the plate from the leading edge to the rear of the section of interest.

Equations for Friction Drag Coefficient C_{Df}. Three separate equations are available for calculating the friction drag coefficient C_{Df}. They are

$$C_{Df} = \frac{1.328}{\sqrt{N_R}} \qquad\qquad \begin{array}{c} \text{Laminar} \\ N_R < 5 \times 10^5 \end{array} \tag{14-5}$$

$$C_{Df} = \frac{0.455}{(\log_{10} N_R)^{2.58}} - \frac{1700}{N_R} \quad \begin{array}{c} \text{Transitional} \\ 5 \times 10^5 < N_R < 10^7 \end{array} \tag{14-6}$$

$$C_{Df} = \frac{0.455}{(\log_{10} N_R)^{2.58}} \qquad\qquad \begin{array}{c} \text{Turbulent} \\ N_R > 10^7 \end{array} \tag{14-7}$$

The values of the Reynolds number ranges given in the above equation are approximate but agree reasonably well with experimental results. As stated before, the roughness of the surface near the leading edge of the body has a marked effect on the point where the boundary layer changes from laminar to transitional. Flow in the transitional region has been observed at a Reynolds number as low as 3×10^5.

After determining the friction drag coefficient from Eq. (14–5), (14–6), or (14–7) the magnitude of the friction drag force can be calculated from

$$F_{Df} = C_{Df} \left(\tfrac{1}{2}\rho v^2\right)A \qquad\qquad \textbf{(14–8)}$$

In general the term A refers to the surface area of the body. This would be the case for friction drag on submarine hulls, railroad cars, truck trailer bodies, and similar items.

The procedure for calculating the friction drag force on a body is outlined below.

1. Calculate the Reynolds number for the motion of the body relative to the fluid using Eq. (14–4).
2. Calculate the friction drag coefficient from Eq. (14–5), (14–6), or (14–7).
3. Calculate the surface area of interest.
4. Calculate the drag force using Eq. (14–8).

Programmed Example Problem

Example Problem 14–1. A high speed rail passenger car has a shape similar to that shown in Fig. 14–3. Determine the friction drag force

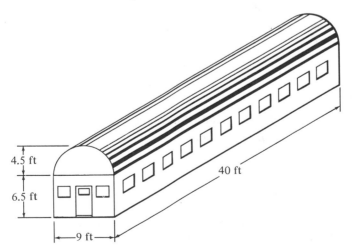

FIGURE 14–3

on the sides and top of the car as it travels 120 miles per hour through still air. The air is at atmospheric pressure, has a temperature of 70°F, and a specific weight of 0.075 lb/ft³.

To begin the solution procedure outlined above, the length L is required for the Reynolds number. Which value should we use?

Considering the sides and top to approximate a flat surface, the length of the car, 40 feet, is used for L. Even though the top of the car is cylindrical in shape, its broad radius allows the flat plate data to be used with satisfactory results.

Now, using Fig. 6–8 to evaluate the dynamic visocity μ, complete the calculation of the Reynolds number.

The result is $N_R = 4.31 \times 10^7$. Here is how this value was found. From Eq. (14–4),

$$N_R = \frac{\rho v L}{\mu}$$

The velocity v of the car relative to the air is

$$v = \frac{120 \text{ mi}}{\text{hr}} \cdot \frac{5280 \text{ ft}}{\text{mi}} \cdot \frac{1 \text{ hr}}{3600 \text{ sec}} = 176 \text{ ft/sec}$$

From Fig. 6–8, $\mu = 3.8 \times 10^{-7}$ lb-sec/ft². The density can be found from the specific weight.

$$\rho = \gamma/g$$
$$\rho = \frac{0.075 \text{ lb}}{\text{ft}^3} \cdot \frac{\sec^2}{32.2 \text{ ft}} = 0.00233 \text{ lb-sec}^2/\text{ft}^4$$

Then,

$$N_R = \frac{\rho v L}{\mu} = \frac{(0.00233)(176)(40)}{3.8 \times 10^{-7}} = 4.31 \times 10^7$$

Now do Step 2 of the procedure.

Since $N_R > 10^7$, Eq. (14–7) should be used to calculate the friction drag coefficient. Complete the calculation before looking at the next panel.

You should have $C_{Df} = 0.00240$. Here are the details.

$$C_{Df} = \frac{0.455}{[\log_{10}(4.31 \times 10^7)]^{2.58}} = \frac{0.455}{(7.634)^{2.58}} = \frac{0.455}{190}$$
$$C_{Df} = 0.00240$$

Now perform Step 3.

The correct result is $A = 1086$ ft². For the two flat sides,

$$A_s = (2)(6.5)(40) \text{ ft}^2 = 520 \text{ ft}^2$$

For the semicylindrical top,

$$A_t = (\tfrac{1}{2})(\pi)(9)(40) \text{ ft} = 566 \text{ ft}^2$$

Then the total area is $A = A_s + A_t = 1086$ ft²

Now complete the calculation of the friction drag force.

From Eq. (14–8),

$$F_{Df} = C_{Df}(\tfrac{1}{2}\rho v^2)A$$
$$F_{Df} = (0.00240)(0.5)(0.00233)(176)^2(1086) \text{ lb}$$
$$F_{Df} = 94 \text{ lb}$$

14–4   Pressure Drag

As a fluid stream flows around a body it tends to adhere to the surface for a portion of the length of the body. Then at a certain point, the thin boundary layer separates from the surface causing a turbulent wake to be formed. See Fig. 14–4. The pressure in the wake is significantly

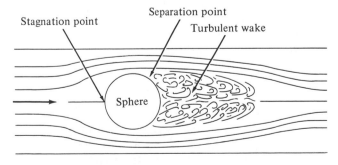

FIGURE 14–4

lower than that at the stagnation point at the front of the body. A net force is thus created which acts in a direction opposite to that of the motion. This force is the pressure drag.

If the point of separation can be caused to occur farther back on the body the size of the wake can be decreased and the pressure drag will be lower. This is the reason for streamlining. Figure 14–5 illustrates the change in the wake caused by the elongation and tapering of the tail of the body. Thus the amount of pressure drag is dependent on the form of the body and the term *form drag* is often used.

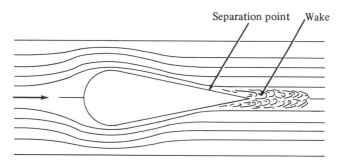

Separation point   Wake

FIGURE 14–5

The pressure drag force is calculated from the equation

$$F_{Dp} = C_{Dp}(\tfrac{1}{2}\rho v^2)A \tag{14–9}$$

in which A is taken to be the maximum cross-sectional area of the body perpendicular to the flow. The coefficient C_{Dp} is the pressure drag coefficient. The total drag is the sum of the friction drag and the pressure drag. That is,

$$F_D = F_{Df} + F_{Dp} \tag{14–10}$$

For different shapes the distribution of the total drag between friction drag and pressure drag varies widely. For a flat plate placed perpendicular to the flow, pressure drag produces nearly all the drag. For a well streamlined shape, friction becomes more significant. As a matter of convenience the term drag will be used to denote pressure drag for the remainder of this section. The drag force will be computed from

$$F_D = C_D(\tfrac{1}{2}\rho v^2)A \tag{14–11}$$

in which C_D is the pressure drag coefficient.

Besides the form of the body, three other parameters have an effect on the pressure drag coefficient. These are the Reynolds number, the surface roughness of the body, and the degree of turbulence in the free fluid stream. Figure 14–6 is a plot of the drag coefficient C_D versus

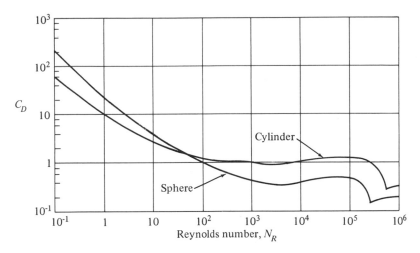

FIGURE 14–6

Drag Coefficients for Spheres and Cylinders

Reynolds number for a smooth sphere in a flow stream having low turbulence. For a sphere, the Reynolds number is based on the diameter D as the characteristic dimension. That is,

$$N_R = \frac{\rho v D}{\mu} = \frac{v D}{\nu} \qquad (14\text{--}12)$$

For very small Reynolds numbers (N_R less than about 1.0) the drag is almost entirely due to friction. The drag force follows the relation

$$F_D = 3 \pi \mu v D \qquad (14\text{--}13)$$

which is called Stoke's law after G. Stokes who developed it in 1851. If Eq. (14–13) is expressed in the form involving a drag coefficient, the resulting value of C_D is $24/N_R$. This plots as a straight line at the left of Fig. 14–6. During this linear portion of the curve, the boundary layer remains attached to the sphere and very little turbulent wake is produced. As the Reynolds number increases, separation occurs and pressure drag becomes more significant.

At a value of the Reynolds number of about 3×10^5 the drag coefficient drops sharply. This is due to the fact that the boundary layer abruptly changes from laminar to turbulent. Concurrently, the point on the sphere at which separation occurs moves farther back, decreasing the size of the wake.

Either roughening the surface or increasing the turbulence in the flow stream can decrease the value of the Reynolds number at which the transition from a laminar to a turbulent boundary layer occurs. This is illustrated in Fig. 14–7. This graph is meant to show typical curve shapes only and should not be used for numerical values.

A cylinder placed with its axis perpendicular to the direction of motion has drag characteristics similar to those of a sphere as shown in Fig. 14–6. Table 14–1 lists values of the drag coefficients for several types of simple shapes at the Reynolds numbers shown. Those shapes for which no Reynolds numbers are shown all have sharp edges which always cause the boundary layer to separate at the same place. There-fore, the drag coefficient for these shapes is nearly independent of Rey-nolds number. This is also substantially true for the triangular cylinders.

14–5   Compressibility Effects and Cavitation

The results reported in the previous section are for conditions in which the compressibility of the fluid (usually air) has little effect on the drag coefficient. These data are valid if the velocity of flow is less than about one-half of the speed of sound in the fluid. Above that speed for air, the character of the flow changes and the drag coefficient increases rapidly.

When the fluid is a liquid such as water, compressibility need not be considered since liquids are very slightly compressible. However, another phenomenon called cavitation must be considered. As the liquid flows

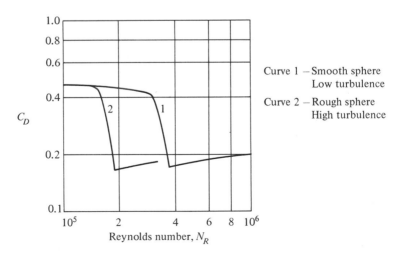

FIGURE 14–7

Effect of Turbulence and Roughness on
C_D *for Spheres*

TABLE 14-1

Typical Drag Coefficients

Shape of Body	Orientation	C_D	N_R
Elliptical cylinders	Flow → (ellipse) 2:1	0.60	4×10^4
		0.47	10^5
		0.35	1.5×10^5
	→ (ellipse) 4:1	0.32	2.5×10^4 to 8×10^4
		0.24	2×10^5
	→ (ellipse) 8:1	0.29	2.5×10^4
		0.20	2×10^5
Square cylinders	→ (square)	1.60	3.5×10^3
		1.90	10^4
		2.05	3×10^4 to 8×10^4
	→ (diamond)	1.60	2×10^4 to 8×10^4
Semitubular cylinders	→ (open facing flow)	1.12	7×10^3 to 4×10^4
	→ (open facing away)	2.30	4×10^4
Triangular cylinders	→ (30° apex facing flow)	1.05	10^4 to 10^5
	→ (30° base facing flow)	1.85	10^4 to 10^5
	→ (60° apex facing flow)	1.39	10^4
	→ (60° base facing flow)	2.20	10^4
	→ (90° apex facing flow)	1.60	10^4
	→ (90° base facing flow)	2.15	10^4

TABLE 14–1 (cont.)

Shape of Body	Orientation	C_D	N_R
Triangular cylinders (cont.)	Flow ⟶ ◁ 120°	1.75	10^4
	⟶ ▷	2.05	10^4
Rectangular plate	a/b		
	1	1.16	
⟶	4	1.17	
b	8	1.23	
⟵ a ⟶	25	1.57	
	50	1.76	
	∞	2.00	
Tandem disks	ℓ/d		
ℓ = spacing	1	0.93	
d = diameter	1.5	0.78	
d	2	1.04	
ℓ	3	1.52	
One circular disk	⟶ d ◯	1.11	
Cylinder	ℓ/d		
ℓ = length	1	0.91	
d = diameter	2	0.85	
d	4	0.87	
ℓ	7	0.99	
Hemispherical cup open back	⟶ ⊂ d	0.41	

TABLE 14–1 (cont.)

Shape of Body	Orientation	C_D	N_R
Hemispherical cup open front	Flow	1.35	
Cone, closed base	$60°$ d	0.51	
	$30°$ d	0.34	

Note: Reynolds numbers shown are based on the length of the body parallel to the flow direction except for the semitubular cylinders for which the characteristic length is the diameter.

Data adapted from references (1) and (7) listed at the end of this chapter.

past a body, the static pressure decreases. If the pressure becomes sufficiently low, the liquid vaporizes forming bubbles. Since the region of low pressure is generally small the bubbles burst when they leave that region. When the collapsing of the vapor bubbles occurs near a surface of the body, rapid erosion or pitting results. Cavitation has other adverse effects when it occurs near control surfaces of boats or on propellers. The bubbles in the water decrease the forces exerted on rudders and control vanes and decrease thrust and performance of propellers.

14–6　　　　Lift and Drag on Airfoils

Lift is defined as a force acting on a body in a direction perpendicular to that of the flow of the fluid. The concepts concerning lift are discussed here with reference to airfoils. The shape of the airfoil comprising the wings of an airplane determine its performance characteristics.

The manner in which an airfoil produces lift when placed in a moving air stream (or moving in still air) is illustrated in Fig. 14–8. As the air flows over the airfoil it achieves a high velocity on the top surface with a corresponding decrease in pressure. At the same time, the pressure on

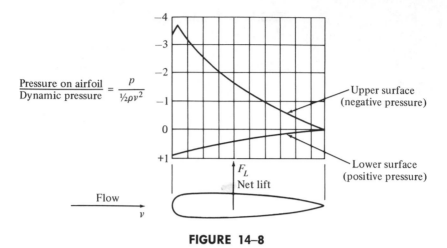

FIGURE 14-8

Pressure Distribution on an Airfoil

the lower surface is increased. The net result is an upward force called lift. It is common to express the lift force F_L as a function of a lift coefficient C_L in a manner similar to that presented for drag.

$$F_L = C_L(\tfrac{1}{2}\rho v^2)A \qquad\qquad (14\text{--}14)$$

The velocity v is the velocity of the free stream of fluid relative to the airfoil. In order to achieve uniformity in the comparison of one shape with another it is customary to define the area A as the product of the span of the wing and the length of the airfoil section called the chord. In Fig. 14–9 the span is b and the chord length is c.

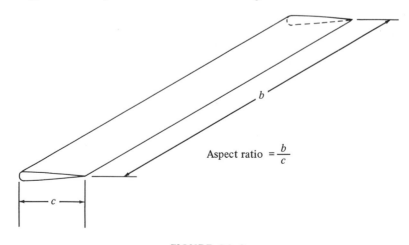

FIGURE 14-9

The value of the lift coefficient C_L is dependent on the shape of the airfoil and also on the angle of attack. Figure 14–10 shows that the angle of attack is the angle between the chord line of the airfoil and the direction of the fluid velocity. Other factors affecting lift are the Reynolds number, the surface roughness, the turbulence of the air stream, the ratio of the velocity of the fluid stream to the speed of sound, and the aspect ratio. Aspect ratio is the name given to the ratio of the span b of the wing to the chord length c. It is important since the characteristics of the flow at the wing tips are different from those toward the center of the span.

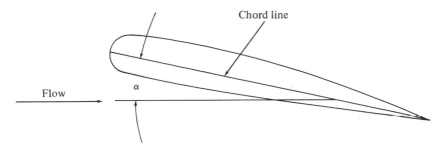

FIGURE 14–10

Angle of Attack, α

The total drag on an airfoil has three components. Friction drag and pressure drag occur as described in Sections 14–3 and 14–4. The third component is called induced drag and it is a function of the lift produced

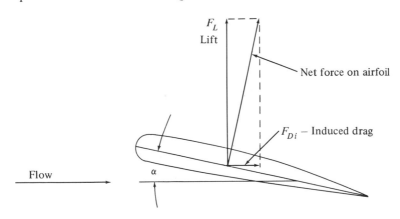

FIGURE 14–11

Induced Drag

by the airfoil. At a particular angle of attack the net resultant force on the airfoil acts essentially perpendicular to the chord line of the section as shown in Fig. 14–11. Resolving this force into vertical and horizontal components produces the true lift force F_L and the induced drag F_{Di}. Expressing the induced drag as a function of a drag coefficient gives

$$F_{Di} = C_{Di}(\tfrac{1}{2}\rho v^2)A \qquad\qquad (14\text{–}15)$$

It can be shown that C_{Di} is related to C_L by the relation

$$C_{Di} = \frac{C_L{}^2}{\pi(b/c)} \qquad\qquad (14\text{–}16)$$

The total drag is then

$$F_D = F_{Df} + F_{Dp} + F_{Di} \qquad\qquad (14\text{–}17)$$

Normally it is the total drag which is of interest in design. A single drag coefficient C_D is determined for the airfoil from which the total drag can be calculated using the relation

$$F_D = C_D(\tfrac{1}{2}\rho v^2)A \qquad\qquad (14\text{–}18)$$

As before, the area A is the product of the span b and the chord length c.

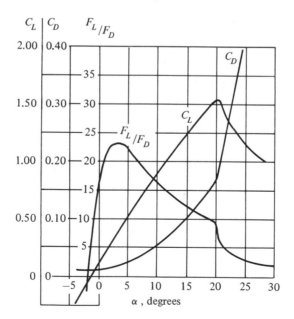

FIGURE 14–12

Airfoil Performance Curves

Two methods are used to present the performance characteristics of airfoil profiles. In Fig. 14–12 the values of C_L, C_D, and the ratio of lift to drag F_L/F_D are all plotted versus the angle of attack as the abscissa. Note that the scale factors are different for each variable. The airfoil to which the data apply has the designation NACA 2409 according to a system established by the National Advisory Committee for Aeronautics. NACA Technical Report 610 explains the code used to describe airfoil profiles. NACA Reports 586, 647, 669, 708, and 824 present the performance characteristics of several airfoil sections.

The second method of presenting data for airfoils is shown in Fig. 14–13. This is called the polar diagram and is constructed by plotting C_L versus C_D with the angle of attack indicated as points on the curve.

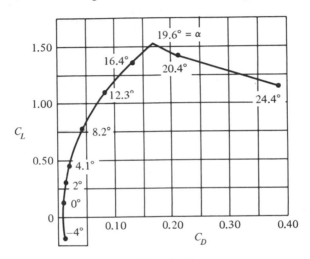

FIGURE 14–13

Airfoil Polar Diagram

In both Fig. 14–12 and Fig. 14–13 it can be seen that the lift coefficient increases with increasing angle of attack up to a point where it abruptly begins to decrease. This point of maximum lift is called the stall point and at this angle of attack the boundary layer of the air stream separates from the upper side of the airfoil. A large turbulent wake is created, greatly increasing drag and decreasing lift.

14–7 **Properties of Air**

In the calculation of the lift and drag forces using Eq. (14–14), (14–15), or (14–18), the density of the air is required. As with all gases the properties of air change drastically with temperature. Also, as altitude above sea level increases the density decreases. Tables 14–2 and 14–3 present the properties of air at various temperatures and altitudes.

TABLE 14–2

Physical Properties of Air at Standard Atmospheric Pressure

Temperature T, °F	Density $\rho \times 10^3$, slugs/ft^3	Specific weight $\gamma \times 10^2$, lb/ft^3	Dynamic viscosity $\mu \times 10^7$, lb-sec/ft^2	Kinematic viscosity $\nu \times 10^4$, ft^2/sec
−40	2.94	9.46	3.12	1.06
−20	2.80	9.03	3.25	1.16
0	2.68	8.62	3.38	1.26
10	2.63	8.46	3.45	1.31
20	2.57	8.27	3.50	1.36
30	2.52	8.11	3.58	1.42
40	2.47	7.94	3.62	1.46
50	2.42	7.79	3.68	1.52
60	2.37	7.63	3.74	1.58
70	2.33	7.50	3.82	1.64
80	2.28	7.35	3.85	1.69
90	2.24	7.23	3.90	1.74
100	2.20	7.09	3.96	1.80
120	2.15	6.84	4.07	1.89
140	2.06	6.63	4.14	2.01
160	1.99	6.41	4.22	2.12
180	1.93	6.21	4.34	2.25
200	1.87	6.02	4.49	2.40
250	1.74	5.60	4.87	2.80

Note: Properties of air for standard conditions at sea level are:

Temperature	59°F
Pressure	2116.2 lb/ft^2
Density	0.00238 slugs/ft^3
Specific weight	0.07648 lb/ft^3
Dynamic viscosity	3.737 $\times$ 10^{-7} lb-sec/ft^2

TABLE 14–3

Properties of the Atmosphere

Altitude ft	Temperature T °F	Pressure p lb/ft^2	Density $\rho \times 10^3$ slugs/ft^3	Specific weight $\gamma \times 10^2$ lb/ft^3	Dynamic viscosity $\mu \times 10^7$ lb-sec/ft^2	Speed of sound ft/sec
0	59.00	2116.2	2.38	7.6475	3.737	1116.4
500	57.22	2078.3	2.34	7.536	3.727	1114.5
1,000	55.43	2040.9	2.25	7.246	3.717	1112.6
5,000	41.17	1760.9	2.05	6.590	3.637	1097.1
10,000	23.36	1455.6	1.76	5.648	3.534	1077.4
20,000	−12.26	973.3	1.27	4.077	3.325	1036.9
30,000	−47.83	629.7	0.889	2.866	3.107	994
40,000	−69.70	393.1	0.585	1.890	2.969	968.1
50,000	−69.70	243.6	0.362	1.171	2.969	968.1
60,000	−69.70	151.0	0.224	0.726	2.969	968.1
70,000	−69.70	93.67	0.140	0.450	2.969	968.1
80,000	−69.70	58.13	0.0873	0.279	2.969	968.1
90,000	−57.20	36.29	0.0544	0.169	3.048	983.4
100,000	−40.89	23.09	0.0339	0.103	3.150	1003.2

References

(1) Baumeister, T., and L. S. Marks. (ed). *Standard Handbook for Mechanical Engineers*. New York: McGraw-Hill Book Co., (7th ed.), 1967.

(2) Daugherty, R. L., and J. B. Franzini. *Fluid Mechanics with Engineering Applications*. New York: McGraw-Hill Book Co., (6th ed.), 1965.

(3) Hoerner, S. F. *Fluid-Dynamic Drag*. Midland Park, N.J.: Published by the author, 1958.

(4) Hurt, H. H., Jr. *Aerodynamics for Naval Aviators*. The Office of the Chief of Naval Operations, U.S. Navy, Document No. NAVWEPS 00-80T-80, 1965.

(5) Jones, B. *Elements of Practical Aerodynamics*. New York: John Wiley & Sons, Inc., (4th ed.), 1950.

(6) Kuethe, A.M., and J. D. Schetzer. *Foundations of Aerodynamics*. New York: John Wiley & Sons, Inc. (2nd ed.), 1959.

(7) Lindsey, W. F. "Drag of Cylinders of Simple Shapes." Report No. 619 — National Advisory Committee for Aeronautics, 1938.

(8) Prandtl, L. *Essentials of Fluid Dynamics*. New York: Hafner Publishing Co., 1952.

(9) Schlichting, H. *Boundary Layer Theory*. New York: McGraw-Hill Book Co., (6th ed.), 1968.

(10) Streeter, V. L. *Fluid Mechanics*. New York: McGraw-Hill Book Co., (5th ed.), 1971.

(11) von Mises, R. *Theory of Flight*. New York: Dover Publications, Inc., 1959. (First published in 1945 by the McGraw-Hill Book Co., New York.)

PRACTICE PROBLEMS

14–1   A cylinder, 1 inch in diameter, is placed perpendicular to a fluid stream having a velocity of 0.5 ft/sec. If the cylinder is 3 feet long, calculate the total drag force if the fluid is (a) water at 60°F, (b) air at 50°F and atmospheric pressure.

14–2   As part of an advertising sign on the top of a tall building, a 3-feet diameter sphere called a "weather ball" glows different colors if the temperature is predicted to drop, rise, or remain about the same. Calculate the force on the weather ball due to winds of 10, 20, 40, and 80 miles per hour if the air is at 40°F.

14–3   Determine the terminal velocity of a 3-inch diameter sphere made of solid aluminum (specific weight = 0.098 lb/in.^{3}) in free fall in (a) castor oil at 77°F, (b) water at 80°F, and (c) air at 80°F and standard atmospheric pressure. Consider the effect of buoyancy.

14–4   Calculate the moment at the base of a flagpole caused by an 80 mile per hour wind. The pole is made of three sections, each 15 feet long, of different size Schedule 80 steel pipe. The bottom section is 6-inch, the middle is 5-inch, and the top is 4-inch. The air is at 40°F and standard atmospheric pressure.

14–5   A pitcher throws a baseball with a velocity of 60 ft/sec. If the ball has a circumference of 9.0 inches, calculate the drag force on the ball in air at 80°F.

14–6   A parachute in the form of a hemispherical cup, 5 feet in diameter is deployed from a car trying for the land speed record. Determine the force exerted on the car if it is moving at 680 miles per hour in air at atmospheric pressure and 80°F.

14–7   Calculate the required diameter of a parachute supporting a man weighing 180 pounds if the terminal velocity in air at 100°F is to be 15 ft/sec.

14–8   A ship tows an instrument in the form of a 30° cone, point first, at 7.5 meters per second in sea water. If the base of the cone has a diameter of 2.20 meters, calculate the force in the cable to which the cone is attached.

14–9   A highway sign is being designed to withstand winds of 80 miles per hour. Calculate the total force on a sign 12 ft × 10 ft if the wind is flowing perpendicular to the face of the sign. The air is at 40°F. Compare the force from this problem with that for problem 13–6. Discuss the reasons for the differences.

14–10   Assuming that a semitrailer behaves as a square cylinder, calculate the force exerted if a 10 mile per hour wind strikes it broadside. The trailer is 8 ft × 8 ft × 40 ft. The air is at 40°F and standard atmospheric pressure.

14–11   A type of level indicator incorporates four hemispherical cups with open fronts mounted as shown in Fig. 14–14. Each cup is 1 inch in diameter. A motor drives the cups at a constant

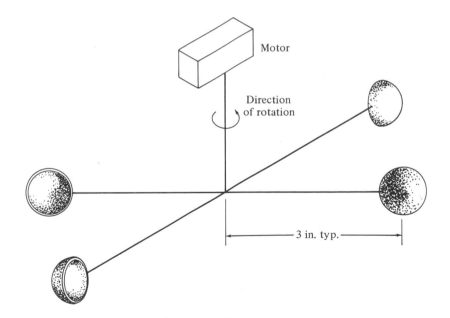

FIGURE 14–14

rotational speed. Calculate the torque which the motor must produce to maintain the motion at 20 revolutions per minute when the cups are in (a) air at 80°F, (b) gasoline at 77°F.

14-12 Determine the wind velocity required to overturn the mobile home sketched in Fig. 14–15 if it is 30 feet long. Consider it to be a square cylinder. The air is at 40°F.

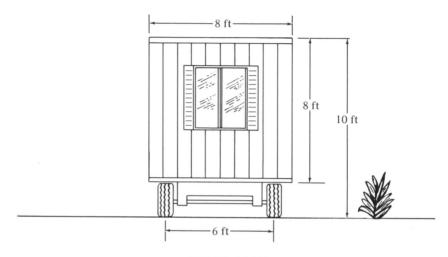

FIGURE 14–15

14-13 A bulk liquid transport truck incorporates a cylindrical tank 6 feet in diameter and 24 feet long. For the tank alone calculate the pressure drag and the friction drag when the truck is traveling at 60 miles per hour in still air at 40°F.

14-14 Calculate the total friction drag on an airfoil whose chord length is 6 feet and whose span is 30 feet. As an approximation, consider it to be a rectangular plate. The airfoil is at 10,000 feet flying at (a) 400 miles per hour, (b) 100 miles per hour.

14-15 For the airfoil whose performance characteristics are shown in Fig. 14–12 determine the lift and drag at an angle of attack of 10°. The airfoil has a chord length of 4.5 feet and a span of 22 feet. Perform the calculation at a speed of 120 miles per hour in the standard atmosphere at (a) 500 feet, (b) 30,000 feet.

14-16 Repeat problem 14–15 if the angle of attack is the stall point, 19.6°.

14-17   For the airfoil in problem 14-15, what load could be lifted from the ground at a takeoff speed of 80 miles per hour when the angle of attack is 15°. The air is at 100°F and standard atmospheric pressure.

14-18   Determine the required wing area for a 3000 pound airplane to cruise at 80 miles per hour if the airfoil is set at an angle of attack of 2.5°. The airfoil has the characteristics shown in Fig. 14-12. The cruise altitude is 20,000 feet in the standard atmosphere.

APPENDIX A

Properties of Water

14.7 psia

Temperature (°F)	Specific Weight γ (lb/ft^3)	Density ρ (slugs/ft^3)	Dynamic Viscosity μ (lb-sec/ft^2)	Kinematic Viscosity ν (ft^2/sec)
32	62.4	1.94	3.66×10^{-5}	1.89×10^{-5}
40	62.4	1.94	3.23×10^{-5}	1.67×10^{-5}
50	62.4	1.94	2.72×10^{-5}	1.40×10^{-5}
60	62.4	1.94	2.35×10^{-5}	1.21×10^{-5}
70	62.3	1.94	2.04×10^{-5}	1.05×10^{-5}
80	62.2	1.93	1.77×10^{-5}	9.15×10^{-6}
90	62.1	1.93	1.60×10^{-5}	8.29×10^{-6}
100	62.0	1.93	1.42×10^{-5}	7.37×10^{-6}
110	61.9	1.92	1.26×10^{-5}	6.55×10^{-6}
120	61.7	1.92	1.14×10^{-5}	5.94×10^{-6}
130	61.5	1.91	1.05×10^{-5}	5.49×10^{-6}
140	61.4	1.91	9.60×10^{-6}	5.03×10^{-6}
150	61.2	1.90	8.90×10^{-6}	4.68×10^{-6}
160	61.0	1.90	8.30×10^{-6}	4.38×10^{-6}
170	60.8	1.89	7.70×10^{-6}	4.07×10^{-6}
180	60.6	1.88	7.23×10^{-6}	3.84×10^{-6}
190	60.4	1.88	6.80×10^{-6}	3.62×10^{-6}
200	60.1	1.87	6.25×10^{-6}	3.35×10^{-6}
212	59.8	1.86	5.89×10^{-6}	3.17×10^{-6}

APPENDIX B

Properties of Common Liquids

14.7 psia and 77°F

	Specific Gravity sg	Specific Weight γ (lb/ft³)	Density ρ (slugs/ft³)	Dynamic Viscosity μ (lb-sec/ft²)
Acetone	0.787	48.98	1.53	6.60×10^{-6}
Alcohol, ethyl	0.787	49.01	1.53	2.29×10^{-5}
Alcohol, methyl	0.789	49.10	1.53	1.17×10^{-5}
Alcohol, propyl	0.802	49.94	1.56	4.01×10^{-5}
Ammonia	0.826	51.41	1.60	—
Benzene	0.876	54.55	1.70	1.26×10^{-5}
Carbon tetrachloride	1.590	98.91	3.08	1.90×10^{-5}
Castor oil	0.960	59.69	1.86	1.36×10^{-2}
Ethylene glycol	1.100	68.47	2.13	3.38×10^{-4}
Gasoline	0.721	45.00	1.40	9.35×10^{-6}
Glycerine	1.263	78.62	2.45	1.98×10^{-2}
Kerosene	0.823	51.20	1.60	3.43×10^{-5}
Linseed oil	0.93	58.0	1.80	6.91×10^{-4}
Mercury	13.633	849.0	26.45	3.20×10^{-5}
Propane	0.495	30.81	0.96	2.30×10^{-6}
Sea water	1.03	64.0	2.00	—
Turpentine	0.87	54.2	1.69	2.87×10^{-5}
Fuel oil, medium	0.852	53.16	1.65	6.25×10^{-5}
Fuel oil, heavy	0.906	56.53	1.76	2.24×10^{-3}

Properties of Petroleum Lubricating Oils

Type or Use	Specific Gravity at 60°F	Kinematic Viscosity, ν				Viscosity Index
		ft²/sec		Saybolt Universal Seconds		
		at 100°F	at 210°F	at 100°F	at 210°F	Index
Automotive Engines						
SAE 10 W	0.870	4.41×10^{-4}	6.46×10^{-5}	190	46	102
20W	0.885	7.64×10^{-4}	9.15×10^{-5}	330	54	96
30	0.891	1.23×10^{-3}	1.22×10^{-4}	530	64	92
40	0.890	1.86×10^{-3}	1.59×10^{-4}	800	77	90
50	0.902	2.91×10^{-3}	2.12×10^{-4}	1250	97	90
10W-30	0.876	8.18×10^{-4}	1.37×10^{-4}	354	69	141
20W-40	0.892	1.00×10^{-3}	1.47×10^{-4}	430	73	135
Gear Units						
SAE 75	0.900	5.06×10^{-4}	7.85×10^{-5}	220	50	121
80	0.934	7.42×10^{-4}	8.50×10^{-5}	320	52	78
90	0.930	3.09×10^{-3}	2.20×10^{-4}	1330	100	91
140	0.937	7.80×10^{-3}	3.66×10^{-4}	3350	160	82
250	0.940	1.31×10^{-2}	5.06×10^{-4}	5660	220	83
Automotive Hydraulic Systems	0.887	4.30×10^{-4}	7.85×10^{-5}	185	50	141

Machine Tool Hydraulic Systems					
Light	0.887	3.44×10^{-4}	5.16×10^{-5}	150	42	64
Medium	0.895	7.21×10^{-4}	7.85×10^{-5}	310	50	66
Heavy	0.901	2.11×10^{-3}	1.51×10^{-4}	910	74	70
Low Temperature	0.844	1.51×10^{-4}	5.60×10^{-5}	74	43	226
Machine Tool Lubricating Oils					
Light	0.881	2.37×10^{-4}	4.20×10^{-5}	105	39	80
Medium	0.915	7.10×10^{-4}	7.53×10^{-5}	305	49	83
Heavy	0.890	2.15×10^{-3}	1.67×10^{-4}	930	80	80

APPENDIX D

Dimensions of Steel Pipe

Schedule 40

Nominal Pipe Size (in.)	Outside Diameter (in.)	Inside Diameter (in.)	Inside Diameter (ft)	Wall Thickness (in.)	Flow Area (in.²)	Flow Area (ft²)
$\frac{1}{8}$	0.405	0.269	0.0224	0.068	0.0568	0.000394
$\frac{1}{4}$	0.540	0.364	0.0303	0.088	0.1041	0.000723
$\frac{3}{8}$	0.675	0.493	0.0411	0.091	0.1910	0.00133
$\frac{1}{2}$	0.840	0.622	0.0518	0.109	0.304	0.00211
$\frac{3}{4}$	1.050	0.824	0.0687	0.113	0.533	0.00370
1	1.315	1.049	0.0874	0.133	0.864	0.00600
$1\frac{1}{4}$	1.660	1.380	0.1150	0.140	1.496	0.01039
$1\frac{1}{2}$	1.900	1.610	0.1342	0.145	2.036	0.01414
2	2.375	2.067	0.1723	0.154	3.36	0.02333
$2\frac{1}{2}$	2.875	2.469	0.2058	0.203	4.79	0.03326
3	3.500	3.068	0.2557	0.216	7.39	0.05132
$3\frac{1}{2}$	4.000	3.548	0.2957	0.226	9.89	0.06868
4	4.500	4.026	0.3355	0.237	12.73	0.08840
5	5.563	5.047	0.4206	0.258	20.01	0.1390
6	6.625	6.065	0.5054	0.280	28.89	0.2006
8	8.625	7.981	0.6651	0.322	50.0	0.3472
10	10.750	10.020	0.8350	0.365	78.9	0.5479
12	12.750	11.938	0.9948	0.406	111.9	0.7771
14	14.000	13.126	1.094	0.437	135.3	0.9396
16	16.000	15.000	1.250	0.500	176.7	1.227
18	18.000	16.876	1.406	0.562	223.7	1.553
20	20.000	18.814	1.568	0.593	278.0	1.931
24	24.000	22.626	1.886	0.687	402	2.792

Schedule 80

Nominal Pipe Size (in.)	Outside Diameter (in.)	Inside Diameter		Wall Thickness (in.)	Flow Area	
		(in.)	(ft)		(in.²)	(ft²)
$\frac{1}{8}$	0.405	0.215	0.01792	0.095	0.0364	0.000253
$\frac{1}{4}$	0.540	0.302	0.02517	0.119	0.0716	0.000497
$\frac{3}{8}$	0.675	0.423	0.03525	0.126	0.1405	0.000976
$\frac{1}{2}$	0.840	0.546	0.04550	0.147	0.2340	0.001625
$\frac{3}{4}$	1.050	0.742	0.06183	0.154	0.432	0.00300
1	1.315	0.957	0.07975	0.179	0.719	0.00499
$1\frac{1}{4}$	1.660	1.278	0.1065	0.191	1.283	0.00891
$1\frac{1}{2}$	1.900	1.500	0.1250	0.200	1.767	0.01227
2	2.375	1.939	0.1616	0.218	2.953	0.02051
$2\frac{1}{2}$	2.875	2.323	0.1936	0.276	4.24	0.02944
3	3.500	2.900	0.2417	0.300	6.61	0.04590
$3\frac{1}{2}$	4.000	3.364	0.2803	0.318	8.89	0.06174
4	4.500	3.826	0.3188	0.337	11.50	0.07986
5	5.563	4.813	0.4011	0.375	18.19	0.1263
6	6.625	5.761	0.4801	0.432	26.07	0.1810
8	8.625	7.625	0.6354	0.500	45.7	0.3174
10	10.750	9.564	0.7970	0.593	71.8	0.4986
12	12.750	11.376	0.9480	0.687	101.6	0.7056
14	14.000	12.500	1.042	0.750	122.7	0.8521
16	16.000	14.314	1.193	0.842	160.9	1.117
18	18.000	16.126	1.344	0.937	204.2	1.418
20	20.000	17.938	1.495	1.031	252.7	1.755
24	24.000	21.564	1.797	1.218	365	2.535

Schedule 160

Nominal Pipe Size (in.)	Outside Diameter (in.)	Inside Diameter		Wall Thickness (in.)	Flow Area	
		(in.)	(ft)		(in.^{2})	(ft^2)
$\frac{1}{8}$	N/A*	—	—	—	—	—
$\frac{1}{4}$	N/A	—	—	—	—	—
$\frac{3}{8}$	N/A	—	—	—	—	—
$\frac{1}{2}$	0.840	0.466	0.03883	0.187	0.1706	0.001185
$\frac{3}{4}$	1.050	0.614	0.05117	0.218	0.2961	0.002056
1	1.315	0.815	0.06792	0.250	0.522	0.003625
$1\frac{1}{4}$	1.660	1.160	0.09667	0.250	1.057	0.007340
$1\frac{1}{2}$	1.900	1.338	0.1115	0.281	1.406	0.009764
2	2.375	1.689	0.1408	0.343	2.240	0.01556
$2\frac{1}{2}$	2.875	2.125	0.1771	0.375	3.55	0.02465
3	3.500	2.626	0.2188	0.437	5.42	0.03764
$3\frac{1}{2}$	N/A	—	—	—	—	—
4	4.500	3.438	0.2865	0.531	9.28	0.06444
5	5.563	4.313	0.3594	0.625	14.61	0.1015
6	6.625	5.189	0.4324	0.718	21.15	0.1469
8	8.625	6.813	0.5678	0.906	36.5	0.2535
10	10.750	8.500	0.7083	1.125	56.7	0.3938
12	12.750	10.126	0.8438	1.312	80.5	0.5590
14	14.000	11.188	0.9323	1.406	98.3	0.6826
16	16.000	12.814	1.068	1.593	129.0	0.8958
18	18.000	14.438	1.203	1.781	163.7	1.137
20	20.000	16.064	1.339	1.968	202.7	1.408
24	24.000	19.314	1.610	2.343	293	2.035

*N/A — Not Available

Dimensions of Steel Tubing

Outside Diameter (in.)	Wall Thickness (in.)	Inside Diameter		Flow Area	
		(in.)	(ft)	(in.²)	(ft²)
$\frac{1}{8}$	0.028	0.069	0.00575	0.00374	2.597×10^{-5}
	0.032	0.061	0.00508	0.00292	2.029×10^{-5}
	0.035	0.055	0.00458	0.00238	1.650×10^{-5}
$\frac{3}{16}$	0.032	0.1235	0.01029	0.01198	8.319×10^{-5}
	0.035	0.1175	0.00979	0.01084	7.530×10^{-5}
$\frac{1}{4}$	0.035	0.180	0.01500	0.02544	1.767×10^{-4}
	0.049	0.152	0.01267	0.01815	1.260×10^{-4}
	0.065	0.120	0.01000	0.01131	7.854×10^{-5}
$\frac{5}{16}$	0.035	0.2425	0.02021	0.04619	3.207×10^{-4}
	0.049	0.2145	0.01788	0.03614	2.509×10^{-4}
	0.065	0.1825	0.01521	0.02616	1.817×10^{-4}
$\frac{3}{8}$	0.035	0.305	0.02542	0.07306	5.074×10^{-4}
	0.049	0.277	0.02308	0.06026	4.185×10^{-4}
	0.065	0.245	0.02042	0.04714	3.274×10^{-4}
$\frac{1}{2}$	0.035	0.430	0.03583	0.1452	1.008×10^{-3}
	0.049	0.402	0.03350	0.1269	8.814×10^{-4}
	0.065	0.370	0.03083	0.1075	7.467×10^{-4}
	0.083	0.334	0.02783	0.0876	6.084×10^{-4}
$\frac{5}{8}$	0.035	0.555	0.04625	0.2419	1.680×10^{-3}
	0.049	0.527	0.04392	0.2181	1.515×10^{-3}
	0.065	0.495	0.04125	0.1924	1.336×10^{-3}
	0.083	0.459	0.03825	0.1655	1.149×10^{-3}
$\frac{3}{4}$	0.049	0.652	0.05433	0.3339	2.319×10^{-3}
	0.065	0.620	0.05167	0.3019	2.097×10^{-3}
	0.083	0.584	0.04867	0.2679	1.860×10^{-3}
	0.109	0.532	0.04433	0.2223	1.544×10^{-3}
$\frac{7}{8}$	0.049	0.777	0.06475	0.4742	3.293×10^{-3}
	0.065	0.745	0.06208	0.4359	3.027×10^{-3}
	0.083	0.709	0.05908	0.3948	2.742×10^{-3}
	0.109	0.657	0.05475	0.3390	2.354×10^{-3}

Outside Diameter (in.)	Wall Thickness (in.)	Inside Diameter		Flow Area	
		(in.)	(ft)	(in.^{2})	(ft^2)
1	0.049	0.902	0.07517	0.6390	4.438×10^{-3}
	0.065	0.870	0.07250	0.5945	4.128×10^{-3}
	0.083	0.834	0.06950	0.5463	3.794×10^{-3}
	0.109	0.782	0.06517	0.4803	3.335×10^{-3}
$1\frac{1}{4}$	0.049	1.152	0.09600	1.0423	7.238×10^{-3}
	0.065	1.120	0.09333	0.9852	6.842×10^{-3}
	0.083	1.084	0.09033	0.9229	6.409×10^{-3}
	0.109	1.032	0.08600	0.8365	5.809×10^{-3}
$1\frac{1}{2}$	0.065	1.370	0.1142	1.474	1.024×10^{-2}
	0.083	1.334	0.1112	1.398	9.706×10^{-3}
	0.109	1.282	0.1068	1.291	8.964×10^{-3}
$1\frac{3}{4}$	0.065	1.620	0.1350	2.061	1.431×10^{-2}
	0.083	1.584	0.1320	1.971	1.368×10^{-2}
	0.109	1.532	0.1277	1.843	1.280×10^{-2}
	0.134	1.482	0.1235	1.725	1.198×10^{-2}
2	0.065	1.870	0.1558	2.746	1.907×10^{-2}
	0.083	1.834	0.1528	2.642	1.835×10^{-2}
	0.109	1.782	0.1485	2.494	1.732×10^{-2}
	0.134	1.732	0.1443	2.356	1.636×10^{-2}

APPENDIX F

Copper Tubing

Type K

Recommended for underground service and general plumbing

Nominal Tube Size (in.)	Outside Diameter (in.)	Wall Thickness (in.)	Inside Diameter		Flow Area	
			(in.)	(ft)	(in.²)	(ft²)
$\frac{1}{4}$	0.375	0.035	0.305	0.02542	0.07306	0.000507
$\frac{3}{8}$	0.500	0.049	0.402	0.03350	0.1269	0.000881
$\frac{1}{2}$	0.625	0.049	0.527	0.04392	0.2181	0.001515
$\frac{5}{8}$	0.750	0.049	0.652	0.05433	0.3339	0.002319
$\frac{3}{4}$	0.875	0.065	0.745	0.06208	0.4359	0.003027
1	1.125	0.065	0.995	0.08292	0.7776	0.005400
$1\frac{1}{4}$	1.375	0.065	1.245	0.1038	1.217	0.008454
$1\frac{1}{2}$	1.625	0.072	1.481	0.1234	1.723	0.01196
2	2.125	0.083	1.959	0.1633	3.014	0.02093
$2\frac{1}{2}$	2.625	0.095	2.435	0.2029	4.657	0.03234
3	3.125	0.109	2.907	0.2423	6.637	0.04609
$3\frac{1}{2}$	3.625	0.120	3.385	0.2821	8.999	0.06249
4	4.125	0.134	3.857	0.3214	11.68	0.08114
5	5.125	0.160	4.805	0.4004	18.13	0.1259
6	6.125	0.192	5.741	0.4784	25.89	0.1798
8	8.125	0.271	7.583	0.6319	45.16	0.3136
10	10.125	0.338	9.449	0.7874	70.12	0.4870
12	12.125	0.405	11.315	0.9429	100.55	0.6983

Type L

Suitable for interior general plumbing

Nominal Tube Size (in.)	Outside Diameter (in.)	Wall Thickness (in.)	Inside Diameter (in.)	Inside Diameter (ft)	Flow Area (in.^{2})	Flow Area (ft^2)
$\frac{1}{4}$	0.375	0.030	0.315	0.02625	0.07793	0.000541
$\frac{3}{8}$	0.500	0.035	0.430	0.03583	0.1452	0.001008
$\frac{1}{2}$	0.625	0.040	0.545	0.04542	0.2333	0.001620
$\frac{5}{8}$	0.750	0.042	0.666	0.05550	0.3484	0.002419
$\frac{3}{4}$	0.875	0.045	0.785	0.06542	0.4840	0.003361
1	1.125	0.050	1.025	0.08542	0.8252	0.005730
$1\frac{1}{4}$	1.375	0.055	1.265	0.1054	1.257	0.008728
$1\frac{1}{2}$	1.625	0.060	1.505	0.1254	1.779	0.01235
2	2.125	0.070	1.985	0.1654	3.095	0.02149
$2\frac{1}{2}$	2.625	0.080	2.465	0.2054	4.772	0.03314
3	3.125	0.090	2.945	0.2454	6.812	0.04730
$3\frac{1}{2}$	3.625	0.100	3.425	0.2854	9.213	0.06398
4	4.125	0.110	3.905	0.3254	11.98	0.08317
5	5.125	0.125	4.875	0.4063	18.67	0.1296
6	6.125	0.140	5.845	0.4871	26.83	0.1863
8	8.125	0.200	7.725	0.6438	46.87	0.3255
10	10.125	0.250	9.625	0.8021	72.76	0.5053
12	12.125	0.280	11.565	0.9638	105.05	0.7295

APPENDIX G

Areas of Circles

Diameter		Area	
(in.)	(ft)	(in.^{2})	(ft^2)
0.25	0.0208	0.0491	0.000341
0.50	0.0417	0.1963	0.00136
0.75	0.0625	0.442	0.00307
1.00	0.0833	0.785	0.00545
1.25	0.1042	1.227	0.00852
1.50	0.1250	1.767	0.01227
1.75	0.1458	2.405	0.01670
2.00	0.1667	3.142	0.0218
2.25	0.1875	3.976	0.0276
2.50	0.2083	4.909	0.0341
2.75	0.229	5.940	0.0412
3.00	0.250	7.069	0.0491
3.50	0.292	9.621	0.0668
4.00	0.333	12.57	0.0873
4.50	0.375	15.90	0.1105
5.00	0.417	19.64	0.1364
6.00	0.500	28.27	0.1964
8.00	0.667	50.27	0.3491
10.00	0.833	78.54	0.545
12.00	1.00	113.1	0.785
18.00	1.50	254.5	1.767
24.00	2.00	452.4	3.142
30.00	2.50	706.9	4.909
36.00	3.00	1018	7.069
42.00	3.50	1385	9.621
48.00	4.00	1810	12.57
54.00	4.50	2290	15.90
60.00	5.00	2827	19.64
72.00	6.00	4072	28.27
84.00	7.00	5542	38.49

APPENDIX H

Conversion Factors

English Gravitational Unit System
to
International System of Units (SI)

Quantity	English Unit	SI Unit	Symbol	Equivalents Units
Length	1 foot (ft) =	0.3048 meter	m	—
Mass	1 slug =	14.59 kilogram	kg	—
Time	1 second (sec) =	1.0 second	s	—
Force	1 pound (lb) =	4.448 newton	N	$kg\text{-}m/s^2$
Pressure	1 lb/ft^2 =	47.88 pascal	Pa	N/m^2 or $kg/m\text{-}s^2$
Energy	1 ft-lb =	1.356 joule	J	N-m or $kg\text{-}m^2/s^2$
Power	1 ft-lb/sec =	1.356 watt	W	J/s

Other Convenient Conversion Factors

Length

  1 foot = 12 inches   1 mile = 1.609 km
  1 foot = 0.3048 meter   1 km = 1000 meters
  1 inch = 2.54 cm   1 meter = 100 cm
  1 mile = 5280 feet   1 meter = 1000 mm

Area

  1 ft^2 = 144 in.^{2}
  1 in.^{2} = 6.452 cm^2
  1 m^2 = 10.76 ft^2

Volume

  1 ft^3 = 7.48 gal   1 liter = 61.02 in.^{3}
  1 ft^3 = 1728 in.^{3} 1 liter = 1000 cm^3
  1 ft^3 = 28.32 liters   1 m^3 = 1000 liters
  1 gal = 231 in.^{3} 1 in.^{3} = 16.39 cm^3

Volume Flow Rate

  1 ft^3/sec = 449 gal/min
  1 ft^3/sec = 0.646 million gal/day
  1 ft^3/min = 0.472 liters/sec

Force

1 lb =	4.448	newtons	1 ton	= 2000 lb	
1 lb =	16	ounces	1 newton	$= 1.0 \times 10^5$ dynes	
1 lb =	7000	grains			

Energy

1 ft-lb	=	1.356 joules
1 BTU	= 778	ft-lb
1 watt-hr	= 2656	ft-lb

Power

1 hp =	550	ft-lb/sec	1 ft-lb/sec = 1.356 watts	
1 hp =	33,000	ft-lb/min	1 ft-lb/sec = 4.626 BTU/hr	
1 hp =	745.7	watts		

Dynamic Viscosity, μ

Units and Conversion Factors

*Multiply number in table by viscosity in given unit
to obtain viscosity in desired unit.*

given unit ＼ desired unit	lb-sec/ft²	Newton-sec / meter²	poise
lb-sec/ft²	1	47.88	478.8
Newton-sec* / meter²	2.089×10^{-2}	1	10
poise**	2.089×10^{-3}	0.1	1
centipoise	2.089×10^{-5}	0.001	0.01

*Standard unit in the International System (SI)
   Equivalent unit: kg/m-sec
**dyne-sec/cm²
   Equivalent unit: gram/cm-sec

Example: Given a viscosity measurement of 200 centipoises, the viscosity
in lb-sec/ft² is:

$$200 \text{ centipoises} \cdot \frac{2.089 \times 10^{-5} \text{ lb-sec/ft}^2}{\text{centipoise}} = 418 \times 10^{-5} \text{ lb-sec/ft}^2$$

Kinematic Viscosity, ν

Units and Conversion Factors

Multiply number in table by viscosity in given unit
to obtain viscosity in desired unit.

given unit ↓ \ desired unit →	ft²/sec	SSU	meter²/sec	stoke
ft²/sec	1	4.29×10^5	9.290×10^{-2}	929.0
SSU*	2.33×10^{-6}	1	2.17×10^{-7}	2.17×10^{-3}
m²/sec**	10.764	4.61×10^6	1	10^4
stoke***	1.076×10^{-3}	4.61×10^2	10^{-4}	1
centistokes	1.076×10^{-5}	4.61	10^{-6}	0.01

*Saybolt Seconds, Universal (conversions approximate for SSU > 100)
   For SSU < 100: $\nu = (0.243\text{SSU} - 210/\text{SSU})(10^{-5})$ ft²/sec
**Standard unit in the International System (SI)
***cm²/sec

Example: Given a viscosity measurement of 200 centistokes, the viscosity
in ft²/sec is:

$$200 \text{ centistokes} \cdot \frac{1.076 \times 10^{-5} \text{ ft}^2/\text{sec}}{\text{centistoke}} = 215 \times 10^{-5} \text{ ft}^2/\text{sec}$$

APPENDIX I

Properties of Areas

Section	Area of Section A	Distance to Centroidal Axis, $\bar{y}$	Moment of Inertia About Centroidal Axis, I_c
Square	H^2	$H/2$	$H^4/12$
Rectangle	BH	$H/2$	$BH^3/12$
Triangle	$BH/2$	$H/3$	$BH^3/36$
Circle	$\pi D^2/4$	$D/2$	$\pi D^4/64$

Section	Area of Section A	Distance to Centroidal Axis, $\bar{y}$	Moment of Inertia About Centroidal Axis, I_c
Ring 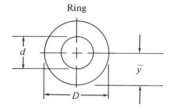	$\dfrac{\pi(D^2 - d^2)}{4}$	$D/2$	$\dfrac{\pi(D^4 - d^4)}{64}$
Semicircle 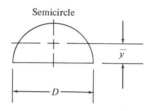	$\pi D^2/8$	$0.212D$	$0.007D^4$
Trapezoid 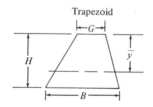	$\dfrac{H(G + B)}{2}$	$\dfrac{H(G + 2B)}{3(G + B)}$	$\dfrac{H^3(G^2 + 4GB + B^2)}{36(G + B)}$

Properties of Solids

Shape	Volume V	Distance to Centroid, $\bar{y}$
Cube	H^3	$H/2$ from any face
Rectangular prism	BHG	$B/2$, $H/2$, or $G/2$ from a particular face
Cylinder	$\dfrac{\pi D^2 H}{4}$	$H/2$
Pyramid	$\dfrac{BGH}{3}$	$H/4$

Shape	Volume V	Distance to Centroid, $\bar{y}$
Hollow cylinder	$\dfrac{\pi H(D^2 - d^2)}{4}$	$H/2$
Sphere	$\dfrac{\pi D^3}{6}$	$D/2$
Hemisphere	$\dfrac{\pi D^3}{12}$	$3D/16$
Cone	$\dfrac{\pi D^2 H}{12}$	$H/4$

APPENDIX K

Notation

Symbol	Explanation
a	Acceleration
A	Area
BM	Distance from center of buoyancy to metacenter
C	Discharge coefficient
C_L	Loss coefficient
cb	Center of buoyancy
cg	Center of gravity
d	Depth of fluid
d	Diameter
D	Diameter
e_M	Mechanical efficiency
E	Energy
f	Friction factor
F	Force
FE	Flow energy
F_b	Buoyant force
F_R	Resultant force due to fluid pressure
g	Acceleration due to gravity (32.2 ft/sec^2)
h	Change in elevation
h	Height of a manometer liquid column
h_L	Energy loss per pound of fluid flowing
h_A	Energy added per pound of fluid flowing
h_R	Energy removed per pound of fluid flowing
H	Height of fluid above a reference line
I	Total moment of inertia of any area about its centroidal axis
I_c	Moment of inertia of a simple area about its centroidal axis
k	Flow resistance coefficient for a pipe in a network
KE	Kinetic energy
ℓ	Distance along a surface
ℓ_c	Distance to the centroid of an area
ℓ_p	Distance to the center of pressure
L	Length
L	Width of an open channel
L	Width of a weir crest
L_e	Equivalent length
m	Mass
mc	Metacenter
M	Mass flow rate
n	Manning's resistance factor for uniform open channel flow

Symbol	Explanation
n	An exponent
N_F	Froude number
N_R	Reynolds number
p	Pressure
p_{abs}	Absolute pressure
p_{atm}	Atmospheric pressure
p_{gage}	Gage pressure
P	Power
P_A	Power added by a pump to the fluid
P_I	Power input to a pump
P_O	Power output from a fluid motor
P_R	Power removed by a fluid motor from the fluid
PE	Potential energy
Q	Volume flow rate
r	Radius
R	Hydraulic radius
R_x	Reaction force in the x-direction
s	Distance or displacement
sg	Specific gravity
S	Slope in open channel flow
t	Time
T	Width of free surface in open channel flow
v	Velocity
V	Volume
w	Weight
W	Weight flow rate
WP	Wetted perimeter
x, y	Coordinates (perpendicular directions)
y	Fluid depth
y_c	Critical depth in open channel flow
y_h	Hydraulic depth in open channel flow
$\bar{y}$	Distance to the centroid of a simple area
$\bar{Y}$	Distance to the centroid of a composite area
z	Elevation
γ	Specific weight
ϵ	Average height of surface roughness
θ	An angle
μ	Dynamic viscosity
ν	Kinematic viscosity
π	Pi = 3.1416
ρ	Density
τ	Shearing stress
ϕ	An angle

Centroid of a Composite Area

The table in Appendix I gives the relationships for calculating the location of the centroid of seven different types of simple areas. Many complex areas can be considered to be made up of two or more of these simple areas and are then called composite areas.

To compute the location of the centroid of a composite area, the following principle is applied. The product of the total area times the distance to the centroid of a composite area is equal to the sum of the products of the area of each component part times the distance to its centroid. This can be expressed as the mathematical equation,

$$A_T \bar{Y} = \Sigma(A\bar{y}) \tag{A-1}$$

where

$\quad A_T =$ total area of the composite area

$\quad \bar{Y} =$ distance from some reference axis to the centroid of the composite area

$\quad A =$ area of a certain component part

$\quad y =$ distance from the reference axis to the centroid of a certain component part

But the total area A_T is equal to the sum of the areas of the component parts. That is,

$$A_T = \Sigma A$$

Then

$$\bar{Y}\Sigma A = \Sigma(A\bar{y}) \tag{A-2}$$

Solving for $\bar{Y}$,

$$\bar{Y} = \frac{\Sigma(A\bar{y})}{\Sigma A} \tag{A-3}$$

This equation is used to calculate the location of the centroidal axis of a composite area relative to a reference axis.

Example Problem Determine the location of the centroid of the composite area shown in Fig. A–1.

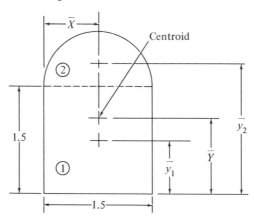

FIGURE A–1

Solution. The area is made up of a square and a semicircle. Calling the square part 1 and the semicircle part 2,

$$A_1 = (1.5 \text{ in.})^2 = 2.25 \text{ in.}^2$$
$$A_2 = \frac{\pi(1.5 \text{ in.})^2}{8} = 0.885 \text{ in.}^2$$

Using the base of the area as the reference axis,

$$\bar{y}_1 = 1.5 \text{ in.}/2 = 0.75 \text{ in.}$$
$$\bar{y}_2 = 1.5 \text{ in.} + 0.212 (1.5 \text{ in.}) = 1.818 \text{ in.}$$

The individual terms in Eq. (A–3) can be calculated in an organized manner by preparing a table like that shown below.

Part	A	$\bar{y}$	$A\bar{y}$
1	2.250 in.^{2}	0.750 in.	1.69 in.^{3}
2	0.885 in.^{2}	1.818 in.	1.61 in.^{3}

$\Sigma A = 3.135$ in.^{2} $\qquad\qquad$ $\Sigma(A\bar{y}) = 3.30$ in.^{3}

Then,

$$\bar{Y} = \frac{\Sigma(A\bar{y})}{\Sigma A}$$

$$\bar{Y} = \frac{3.30 \text{ in.}^3}{3.135 \text{ in.}^2} = 1.05 \text{ in.}$$

Since the area is symmetrical about a vertical axis its centroid lies on the axis of symmetry. Then $\bar{X} = 0.75$ in.

Index